Navigating the Changing Landscape of Formal and Informal Science Learning Opportunities

Deborah Corrigan • Cathy Buntting
Alister Jones • John Loughran
Editors

Navigating the Changing Landscape of Formal and Informal Science Learning Opportunities

 Springer

Editors
Deborah Corrigan
Faculty of Education
Monash University
Clayton, VIC, Australia

Cathy Buntting
University of Waikato
Hamilton, New Zealand

Alister Jones
University of Waikato
Hamilton, New Zealand

John Loughran
Monash University
Clayton, VIC, Australia

ISBN 978-3-030-07850-8   ISBN 978-3-319-89761-5   (eBook)
https://doi.org/10.1007/978-3-319-89761-5

Printed on acid-free paper

This Springer imprint is published by the registered company Springer International Publishing AG part
of Springer Nature.
The registered company address is: Gewerbestrasse 11, 6330 Cham, Switzerland

Preface

This is the fifth book in a series initiated by the Monash University – Kings' College London International Centre for Study of Science and Mathematics Curriculum and in partnership with Waikato University. The Monash-Kings' College Centre was established in 2002 with initial support from the Monash University Research Fund (new areas). The Centre for Science and Technology Education Research at Waikato University and the Centre for Science, Mathematics and Technology Education at Monash University have had a formal partnership agreement since 2003 and have worked cooperatively in many areas.

The first book in the series, *The Re-Emergence of Values in Science Education* (D. Corrigan, J. Dillon & R. Gunstone [Eds.], Rotterdam: Sense Publishers, 2007), considered the state of science education in the twenty-first century through the lens of values. The book presented a 'big picture' of what science education might be like if values once again became central in science education. A decade ago (when this first book was conceptualised) the overwhelming experiences of those who were teaching science were in an environment which had seen the de-emphasising of values fundamentally inherent in both science and science education. There was a disparity between the evolutionary process that science was – and still is – undertaking and that undertaken by science education (and school science education in particular).

In the second book, *The Professional Knowledge Base of Science Teachers* (D. Corrigan, J. Dillon & R. Gunstone [Eds.], Dordrecht, Springer, 2011), our focus was on exploring what expert science education knowledge and practices may look like in the then slowly emerging 'bigger picture' of the re-emergence of values, a focus we saw as a logical step on from the focus on values in the first book. We noted in the Foreword to this first book that the focus of the book was on 'exploring what expert science education knowledge and practices may look like in the emerging "bigger picture" of the re-emergence of values'.

In the third book, *Valuing Assessment in Science education: Pedagogy, Curriculum, Policy* (D. Corrigan, R. Gunstone & A. Jones [Eds.], Dordrecht, Springer, 2013), we took what we saw as the clear logical next step in this sequence of foci begun with our exploration of *The Re- Emergence of Values in Science*

Education; the reality of education is that it is assessment that is almost always the strongest force shaping implemented curriculum, teacher development and behaviour, student approaches to learning, etc. This book considered the 'big picture' of assessment in science education, from the strategic/policy to individual classroom levels. While some classroom case studies were presented, they focused more on teachers than students, and so considered assessment more in terms of what teachers plan and do than in terms of the impacts on students.

This fourth book, *The Future of Learning Science: What's in It for the Learner?* (D.Corrigan, C.Buntting, J.Dillon, A.Jones & R.Gunstone [Eds.], Dordrcht, Springer, 2015), considers learning – the forms of science that better represent the nature of science in the twenty-first century, the purposes we might adopt for the learning of school science, the forms this learning might better take, and how this learning happens (with particular concern for the need to better engage students with their school science and the need to place the burgeoning range of digital technologies into a more informed context than the narrow and uncritical contexts in which these are too commonly considered). An important overarching theme we seek is to represent and value the perspective of the learner.

The fifth book moves on again from *Re-emergence of Values/Professional Knowledge Base/Assessment /Learners* to consider learning science as a fundamental issue in research in science education, in curriculum development and implementation in science education as well as in science teaching and learning. This book takes a broad and deep view of research involving learning opportunities that are afforded to learners of science when the focus is on linking the formal and informal science education sectors. We use the metaphor of a "landscape" as it emphasises how we see the possible movement within a landscape that is inclusive of formal, informal and free-choice opportunities (rather than the not uncommon formal sector assumption that the informal sector should somehow serve the formal and free choice is not part of education at all). This book explores opportunities for seeking to change formal school science education via the perspectives and achievements of the informal and free-choice science education sector as part of the education landscape.

We used the same approach to the creation of this fifth book as we did with the previous four. In a desire to achieve in this edited collection both the creation of a cohesive contribution to the literature and having authors able to assert their own voices without restrictive briefs from us as editors, we again organised a workshop involving the authors and ourselves to enable a more interactive and formative writing process. Authors completed a first draft of their chapters in time to distribute them to all workshop participants before we met. The workshop then involved intensive discussions of individual chapters and feedback to authors, and considerations of the overall structure and cohesion of the volume. Authors then rewrote their chapters in the light of these forms of feedback. As with the previous books, the workshop was scheduled around the European Science Education Research Association (ESERA) conference and took place at the Monash University Centre in Prato (Italy).

This procedure had previously been used very successfully in the production of two other books in which the editors had variously been involved [P. Fensham, R. Gunstone & R. White (Eds.) (1994). *The content of science: A constructivist approach to its teaching and learning*. London: Falmer Press; R. Millar, J. Leach & J. Osborne (Eds.) (2000) *Improving science education: The contributions of research*. Milton Keynes: Open University Press], and has been more recently adopted by other science education researchers. We believe that this process significantly improves the quality of the final product and provides an opportunity for what is sadly a very rare form of professional development – considered and formative and highly collaborative (and totally open) discussions of one's work by one's peers.

We gratefully acknowledge the funding of the workshop through contributions from Monash University and Waikato University, and the commitment, openness and sharing of the participants in the workshop – all authors and editors – that shaped the book.

Clayton, VIC, Australia   Deborah Corrigan
Hamilton, New Zealand   Cathy Buntting
Hamilton, New Zealand   Alister Jones
Clayton, VIC, Australia   John Loughran
July 2017

Contents

About the Authors

Steve Alsop is a Professor in the Faculty of Education, York University, Canada. He teaches and supervises undergraduate and graduate students in the fields of education, science and technology studies, environmental sustainability and interdisciplinary studies. His research explores the personal, social, political and pedagogical articulations of scientific knowledge and technologies in educational settings and contexts. Such settings include the public sphere, cultural institutions (museums and science centres), new social movements, schools and universities. His teaching, research and writing are informed by a commitment and belief in the importance and hopes of knowledge and education building a more wondrous, humane, diverse, equitable and just world.

Cathy Buntting is a Senior Research Fellow at the University of Waikato, New Zealand. She has a Masters degree in biochemistry and a PhD in science education, and is Director of the Science Learning Hub, a significant online portal linking the science and education sectors. Recent research projects have included the development of students' futures thinking and the role of ICTs in transforming science learning and teaching.

Deborah Corrigan is a Professor of Science Education and Deputy Dean of the Faculty of Education at Monash University. After working as a chemistry and biology teacher, she worked at Monash University in chemistry and science education, particularly in teacher preparation. Her research interests include industry and technology links with science, curriculum design, science and STEM education policy and the values that underpin science education. However, her main research interest remains improving the quality of chemistry and science education so that it is relevant to students.

Bronwen Cowie is Professor and Director of the Wilf Malcolm Institute of Educational Research, the University of Waikato, New Zealand. Bronwen has extensive experience in classroom-based research using interviews, observations, and the collection of student work and video as a data generation tools. She is a co-director of the Science Learning Hub, a Ministry of Business, Innovation and Employment initiative to make New Zealand science accessible to New Zealand teachers via a multimedia web-based resource. Her research interests include the links between assessment for learning [AfL] and culturally responsive pedagogy, and the use of ICTs in STEM education from primary to tertiary.

Suzanne Deefholts holds a Bachelor of Education from Monash University and a Master of Education (Student Wellbeing) from Melbourne University. She is passionate about all areas of the curriculum with a particular emphasis on science education. Suzanne has been a primary teacher for 11 years, much of this time teaching at St. Joseph's Primary School, Crib Point Victoria. During her career Suzanne has worked in various leadership roles including Science and Sustainability Leader, Religious Education Leader, Wellbeing Leader, Graduate Teacher Coach and as a Facilitator for the very successful teacher professional learning program *Contemporary Approaches to Primary Science Program*, an initiative of Catholic Education Melbourne. She is presently co-teaching Grade 5/6 at St. Josephs and part of her role involves working with students in the school's purpose built, student operated/teacher facilitated café. Joey's Café is both a contemporary learning and community outreach.

Lynn D. Dierking is Professor, Free-Choice/Informal STEM Learning, Oregon State University, and Director of Strategy & Partnerships, Institute for Learning Innovation. Her research focuses on free-choice, out-of-school learning (in after-school, home- and community-based contexts), with youth and families, particularly those living in poverty and/or not historically engaged in STEM learning across their lifetime. Dierking is PI of a US-NSF project, *SYNERGIES: Customizing Interventions to Sustain Youth STEM Interest and Participation Pathways*, studying and maintaining youths' STEM interest and participation in an under-resourced community by taking an ecosystem approach and is co-PI of a US-NSF/ UK-Wellcome Trust Science Learning+ Partnership project, *Equitable STEM Pathways*. Lynn is on Editorial Boards of *Connected Science Learning*, *Afterschool Matters* and *Journal of Museum Management and Curatorship*. Awards include 2010 American Alliance of Museums' John Cotton Dana Award for Leadership and 2016 NARST (international organization supporting research on science learning) Distinguished Contributions to Science Education Through Research Award.

Justin Dillon is a Professor of Science and Environmental Education at the University of Bristol. After taking a degree in chemistry from Birmingham University, Justin trained as a teacher at Chelsea College and went on to teach in six secondary schools in London. His research originally focused on teaching and learning about chemistry in England and Spain. Over the past 15 years he has focused more on science learning outside the classroom particularly in museums, science centres and botanic gardens in the UK, Europe and elsewhere. Together with two colleagues at King's, Justin co-ordinated the ESRC's Targeted Initiative on Science and Mathematics Education (TISME) and he was a member of the highly influential ASPIRES project. Justin has co-edited a number of books including the

International Handbook of Research on Environmental Education. He was given 'The Outstanding Contributions to Research in Environmental Education Award' by the North American Association for Environmental Education 2013.

John H. Falk is Director of the Institute for Learning Innovation and Sea Grant Professor Emeritus of Free-Choice Learning at Oregon State University. He is internationally known as a leading expert on free-choice learning, the learning that occurs while visiting zoos, aquariums, museums or parks, watching educational television or surfing the Internet for information; he has authored over 200 articles and chapters, two dozen books, and helped to create several nationally important out-of-school educational curricula. His recent work focuses on studying the impacts of zoos, aquariums and science museums on the public's understanding of, interest in and engagement with science and understanding why people utilize free-choice learning settings during their leisure time. Awards include 2010 American Alliance of Museums' John Cotton Dana Award for Leadership, 2013 Council of Scientific Society Presidents Award for Educational Research and 2016 NARST (international organization supporting research on science learning) Distinguished Contributions to Science Education Through Research Award.

Peter J. Fensham is Emeritus Professor of Science Education at Monash University. His current interests are curriculum policy, assessment, and issues associated with the public understanding of science.

Angela Fitzgerald is a Senior Lecturer in the Faculty of Education, Monash University, and the Director of Professional Experience. Her main focus is engaging in activities that support pre- and in-service teachers in developing their confidence and competence in the learning and teaching of science in primary school settings.

Sue Jackson studied at Australian Catholic University, and has since taught across all stages of the primary years. She has worked in Catholic education for 15 years and has had a number of opportunities to mentor teachers through her various leadership roles, including I.C.T Leader, Science Leader, Learning and Teaching Leader and STEM Leader. Susanne is currently the Mathematics Leader, Visible Coach and Deputy Principal at St Joseph's School, Crib Point. She is committed to ongoing professional learning and has greatly benefitted from a range of experiences, including the Primary Mathematics Leadership Program (Monash University), Growth Coaching, Ed Partnerships – Connecting Learning in Communities (CLIC) and Contemporary Approaches to Primary Science. She is passionate about learning through inquiry and facilitating students to build the knowledge, skills and dispositions they need to be successful lifelong learners.

Alister Jones is a Research Professor and Senior Deputy Vice-Chancellor at the University of Waikato, New Zealand. He is Director of the New Zealand Science and Biotechnology Learning Hubs, as well as Director for a number of companies. He has been consulted on educational development in New Zealand, Australia, the UK, the USA, Hong Kong, Chile and Thailand. His research interests span technology and science education, teacher education, curriculum, assessment and educational leadership.

Elaine Khoo a Senior Research Fellow at the Wilf Malcolm Institute of Educational Research, the University of Waikato, New Zealand. Elaine's research interests include teaching and learning in information and communication technology (ICT) supported learning environments, online learning settings with a particular interest in online learning communities, participatory learning cultures and collaborative research contexts. Elaine has been involved in a number of Ministry of Education-funded research projects associated with online learning, Web 2.0 tools and ICTs across the compulsory schooling sector and at tertiary level. Her completed projects investigated networked science inquiry in secondary classrooms, explored the educational affordances of iPads among pre-schoolers and investigated the impact of flipped classrooms in engineering education.

Gillian Kidman is an Associate Professor of Science Education at Monash University. She has research and teaching interests in the sciences and is particularly interested in inquiry forms of teaching and learning and the potential inquiry has for the integration of science with other disciplines. Gillian is currently working on the interpretation of the Inquiry strands of the Australian Curriculum in Science and the Humanities. Her work in the area of disciplinarity and educational inquiry with teachers is providing a fertile ground for the exploration of Inquiry based learning, teaching and skill acquisition. Gillian's research, teaching and curriculum design is award winning at both the State and National levels, and she was a writer and senior advisor of the Australian Curriculum: Science, and Australian Curriculum: Biology.

Greg Lancaster has extensive experience in teaching secondary science and senior physics while undertaking a variety of professional roles in secondary schools in Victoria. For a number of years he worked part-time in the tertiary and secondary sectors as a science teacher and science teacher educator, which afforded him with rich insights into the changing nature of teacher professional practice, the challenges of curriculum design and authentic teacher professional learning. More recently he completed a Masters of Teaching and continues to work part-time for the Faculty of Education at Monash University across a number of roles, including the faculty academic liaison to the John Monash Science School, a science education researcher, and lecturer in physics education. His research interests include the development of conceptual understanding in physics, inquiry-based learning and science teacher professional learning.

Simon Lindsay is the Manager of Improved Learning Outcomes at Catholic Education Melbourne. Simon has worked in Catholic education for 16 years supporting teacher professional learning across a large system of 360 schools. Simon previously held education roles at Swinburne University and Museum Victoria, and started his career as a population ecologist in Santa Clara, California. He has authored numerous publications in science education and has a keen interest in the development of scientific literacy within the broader society.

John Loughran is the Foundation Chair in Curriculum & Pedagogy and Dean of the Faculty of Education, Monash University. John was a science teacher for 10 years before moving into teacher education. His research interests include the fields of teacher education and science education. John was the co-founding editor of *Studying Teacher Education* and is an Executive Editor for *Teachers and Teaching: Theory and Practice* and on the International Editorial Advisory Board for a number of journals including *Teacher Education Quarterly*, *Journal of Reflective Practice* and the *Asia Pacific Forum for Science Teaching and Learning*.

Karen Marangio is a Lecturer at Monash University and an experienced secondary school teacher and tertiary teacher educator, having taught in both contexts for over 15 years. She has extensive experience working on secondary school curriculum and examination panels and has worked in 'informal' environmental education settings. Karen is enjoying the shift into a teaching and research role at Monash University, including teaching within the science, chemistry and psychology education units. Karen's interests revolve around curriculum and pedagogy, including developing the pedagogy of psychology and science pre-service teachers, and the teaching of concepts with science practices to facilitate deeper learning of science, including the value, complexity and relevance of science within our society.

Jasper Montana is a Post-doctoral Fellow in the Department of Politics at Sheffield University, England. His doctoral study involved the processes involved in developing a United Nations Policy on Biodiversity, and he is now associated with Professor James Wilsdon who was the Founding Director of the Science Policy Centre for the Royal Society.

Ana Maria Navas-Iannini is a PhD candidate at the Ontario Institute for Studies in Education at the University of Toronto. Currently, she is developing a comparative study about critical science exhibitions in Brazil and Canada through the lenses of scientific literacy and science communication. Her research interests include science education and science communication in museums and science centres, public engagement with science and technology and scientific literacy in informal educational settings.

Debra Panizzon is Associate Professor of Science Education at Monash University, having held previous academic positions at Flinders University and the University of New England. Prior to commencement in academia she taught junior science in secondary schools along with senior biology. Debra is an experienced science education academic and has worked with both primary and secondary pre-service teachers. Her research interests lie in the areas of cognition, student acquisition of scientific concepts, assessment, and rural and regional education given extensive experience living and working in rural NSW. Importantly, much of her research has emerged from partnerships with science and mathematics teachers, ensuring that theory and practice are inextricably linked.

Erminia Pedretti is Professor of Science Education at the Ontario Institute for Studies in Education (OISE) at the University of Toronto. She is the former director of the Centre for Science, Mathematics and Technology Education located in the Department of Curriculum, Teaching and Learning. Dr. Pedretti teaches in the initial teacher education and graduate programs. Her research focuses on science education in school and non-school settings, environmental education, and teaching and learning about science, technology, society and environment (STSE). She has published over 45 articles, 5 books and 2 teacher education textbooks for primary/ junior and intermediate/senior science methods curriculum and instruction courses. Over the years she has received numerous grants to support her research and graduate students. Her current research grant *Engaging the Public with Controversial Exhibitions at Science Museums* explores, through a series of case studies, controversial and/or issues-based exhibitions and the interface between science communication and visitor engagement.

Léonie Rennie is Emeritus Professor in Science and Technology Education at Curtin University in Perth, Western Australia. Léonie's research interests concern the processes, outcomes and assessment of learning in science and technology, particularly in out-of-school settings, which she has been researching for over two decades. Her scholarly publications include over 200 refereed journal articles, book chapters and monographs, including as co-author, *Knowledge That Counts in a Global Community: Exploring the Contribution of Integrated Curriculum* (2012), and co-editor, *Integrating Science, Technology, Engineering and Mathematics: Issues, Reflections, and Ways Forward* (2012). In 2009, she received the Distinguished Contributions to Science Education Through Research Award from the US-based National Association for Research in Science Teaching.

Nicole Sadler is a practising teacher with 32 years experience teaching primary, secondary and tertiary students in a range of disciplines including science, mathematics, sustainability, music and English. She has worked as a curriculum consultant for Catholic education in both Victoria and Western Australia and she is particularly passionate about integrating indigenous perspectives and sustainability throughout the primary curriculum. Nicole has two Masters degrees and is currently head of science and sustainability and year 1/2 classroom teacher at a small coastal primary school, St Aloysius, in Victoria, Australia.

Alan Smith has extensive experience working in early childhood centres and working in Catholic parish primary schools teaching at all levels of primary education. He is presently the Principal of Holy Child Primary School in Dallas, Victoria, and has been in this role for 8 years. During this time, much of his work has focused on building community relationships and enabling teaching staff to develop an appreciation of the importance that effective pedagogical approaches play in supporting the learning needs of students from a diverse range of EAL backgrounds. Alan's vision is to provide opportunities for families and children to work in safe and inspiring learning environments with access to learning experiences which will enhance their life chances.

Kathy Smith is Senior Lecturer at Monash University. She has expertise in primary science education and a particular research interest in teacher professional learning and the conditions that build teacher capacity for self-directed learning. Kathy began her career as a primary teacher and worked in this role for over 9 years before branching into other areas of education. In recent years, while working as an independent education consultant, Kathy worked across education sectors in projects related to science education and teacher professional learning and assisted with the development and implementation of a sector wide Science Education Strategy for Catholic Education Melbourne. Kathy continues to work closely with teachers in schools in the areas of curriculum planning and STEM education.

Susan M. Stocklmayer was a Professor of Science Communication at Australian National University and the Director of the Centre for the Public Awareness of Science from 1998 to 2015. As part of the University's outreach programs, she has presented festival science shows, lectures and workshops on all five continents. She was awarded an Australian Medal in 2004 for science communication initiatives, and an Australian Order for science communication and science education in 2016. Sue thinks that science communication is the best possible mixture, combining science, theatre (a lifelong interest), multicultural and gender issues and a host of other things at the interface between science and the public.

Grady Venville is Professor and Dean of Coursework Studies at the University of Western Australia. She is responsible for the quality and structural integrity of the curriculum across all coursework within the university. Her research interests focus on curriculum integration, conceptual change and cognitive acceleration. Grady is widely published including as co-editor (with Vaille Dawson) of the popular textbook *The Art of Teaching Science* (Allen & Unwin, 2012) and co-author (with Léonie Rennie and John Wallace) of *Knowledge That Counts in a Global Community: Exploring the Contribution of Integrated Curriculum* (Routledge, 2012).

John Wallace is a Professor at the Ontario Institute for Studies in Education, University of Toronto with a 45 year career in education, including work in classrooms, schools and school systems. His teaching and research interests include science teaching, teacher learning, teacher knowledge, curriculum integration and qualitative inquiry. He is Editor-in-Chief of the *Canadian Journal of Science, Mathematics and Technology Education*. John's recent co-edited/authored books include *Contemporary Qualitative Research: Exemplars for Science and Mathematics Educators* (Springer, 2007, with Peter Taylor) and *Knowledge That Counts in a Global Community: Exploring the Contribution of Integrated Curriculum* (Routledge, 2012, with Léonie Rennie and Grady Venville).

Navigating the Changing Landscape of Formal and Informal Science Learning Opportunities

Deborah Corrigan, Cathy Buntting, Alister Jones, and John Loughran

Abstract A Google Scholar search of outreach, informal, non-formal, and out-of-school science education highlights a prolific and rich area of experiences offered by education institutions (e.g., universities), educational centres (e.g., zoos, museums), libraries, the broadcast media, workplaces and other community-based organisations to enhance individual understandings of science (Falk JH, Dierking LD, Museum experience revisited. Left Coast Press, Walnut Creek, 2012).

This book takes a broad and deep view of research involving learning opportunities that are afforded to learners of science when the focus is on linking the formal and informal science education sectors. We use the metaphor of a 'landscape' to emphasise the possible movement and interactions within a landscape that is inclusive of formal, informal and free-choice learning opportunities (rather than the not uncommon formal sector assumption that the informal sector should somehow serve the formal, and that free choice is not part of education at all). This book explores opportunities to change formal school science education by considering some perspectives and achievements from the informal and free-choice science education sector within the wider lifelong, life-wide education landscape. Additionally it explores how science learning that occurs in a more inclusive landscape can demonstrate the potential power of these opportunities to address issues of relevance and engagement that currently plague the learning of science in school settings.

Keywords Informal science learning · Formal science learning · Out-of-school science education

D. Corrigan (✉) · J. Loughran
Monash University, Clayton, VIC, Australia
e-mail: deborah.corrigan@monash.edu; john.loughran@monash.edu

C. Buntting · A. Jones
University of Waikato, Hamilton, New Zealand
e-mail: BUNTTING@waikato.ac.nz; a.jones@waikato.ac.nz

1

A Google Scholar search of outreach, informal, non-formal, and out-of-school science education highlights a prolific and rich area of experiences offered by education institutions (e.g., universities), educational centres (e.g., zoos, museums), libraries, the broadcast media, workplaces and other community-based organisations to enhance individual understandings of science (Falk & Dierking, 2012). While there are nuances identifiable in the way these terms are used in the literature they essentially refer to:

> Activities that occur outside of the school setting, are not developed primarily for school use, are not developed to be part of an ongoing school curriculum, and are characterized as voluntary as opposed to mandatory participation. (Crane, Nicholson, Chen, & Bitgood, 1994, p. 3)

Over time the out-of-school science arena has characterised a major shift in the role and purpose of these educational facilities and institutions (e.g., zoos, museums, and botanical gardens). This has often been in response to greater community interest and awareness as well as a rapidly changing global environment. For example, zoos initially provided a menagerie of animals from all over the world as 'curiosities' with interest largely on anatomy and physiology (Conway, 1969). Today zoos fulfil a more complex role as research, teaching and conservation institutions in their own right, providing protected environments for small pockets of endangered and threatened species while contributing to global captive breeding progammes. In addition to their wildlife conservancy role they strive to educate and establish affective connections with the general public by contextualising scientific knowledge and processes through conveying information about animal adaptations and habitats (Anderson, Kelling, Pressley-Keogh, Bloomsmith, & Mapple, 2003). In so doing they challenge public thinking and personal values, encouraging cultural shifts so that critical issues like sustainability at both a local and global level become the responsibility of us all. Importantly, the chance to engage with this type of outreach has been further expanded with the provision of online resources, 'virtual' experiences and interactive features that can be accessed by anyone, any time, from any device connected to the Internet.

This book takes a broad and deep view of research involving learning opportunities that are afforded to learners of science when the focus is on linking the formal and informal science education sectors. We use the metaphor of a 'landscape' to emphasise the possible movement and interactions within a landscape that is inclusive of formal, informal and free-choice learning opportunities (rather than the not uncommon formal sector assumption that the informal sector should somehow serve the formal, and that free choice is not part of education at all). This book explores opportunities to change formal school science education by considering some perspectives and achievements from the informal and free-choice science education sector within the wider lifelong, life-wide education landscape. Additionally it explores how science learning that occurs in a more inclusive landscape can demonstrate the potential power of these opportunities to address issues of relevance and engagement that currently plague the learning of science in school settings.

Some of the Persistent Issues in Formal Science Education

There are some persistent challenges and issues in science education that need to be addressed. These include:

- a continued highly conservative approach to determining *what* science is included in the school curriculum and the great difficulty many experience in embedding contemporary science in the school curriculum,
- the continued reductionist and simplistic views of the nature of science itself, which unfortunately remains common in both the intended curriculum and actual classroom interactions,
- the continued low levels of student engagement with school science, particularly as schooling progresses,
- the continued (and sad) contrasts between the ways the learning of science is generally conceived and perceived in the formal and informal sectors,
- the different ways in which the learning of science ideas is valued and rewarded in the two sectors,
- the lack of recognition for the transformative impact on many of the above issues that outreach programs (e.g., bringing scientists and school students together) can have in a supportive environment, and
- the slow recognition of the range of potentials (and limitations) offered by digital technologies for changing these issues.

Crucial to challenging these issues is broadening ideas about what science is worthwhile for school students to learn for their everyday lives and their futures. Science learning that is embedded in communities and workplaces must also be considered as worthwhile learning. Hence, the place where science learning takes place is also an important part of broadening of ideas that can challenge these issues.

Additionally, if we are to optimise science learning across an inclusive landscape there is a need to build and develop connections across this landscape (Falk et al., 2015). Such connections will be of a diverse nature, as outlined by the chapters within this book. Partnerships will be fundamentally important and relationships conducive to a shared (and broad) understanding of science learning will be needed. At other times interventions may be needed. Decisions about the relationships that are needed will need to take into account context, voice and boundaries.

Context in this instance refers not only to where the learning will occur, but why such learning is desired, and where it will be applied. Hence out-of-school contexts are fundamentally important as use of what we have learnt is almost always undertaken outside the school and institutional setting. Just as there will be multiple, unique and overlapping contexts, there is a diversity of voices across the wide range of different stakeholders in science learning in both in and out of school contexts – learners, teachers, institutional leaders and managers, funders, curriculum developers, and so on. Importantly the voice of the learner must be a driving force, particularly if we are to address issues of relevance—learners must be able to see purpose in what they are learning. Third, the notion of boundaries is important for successfully navigating the informal and formal science education landscape—recognising the existence of boundaries is a pre-requisite to negotiating their movement. Currently there are numerous boundaries for school learners: what is valued in school as opposed to their lives outside of school, what counts as success in learning school science, knowing when to use which lenses for meaning making, how to build science competence that encompasses a range of real-life contexts, and so on. In simple terms, there are many examples of science education beyond the school that link back into formal education and demonstrate the potential power of navigating the science learning landscape more creatively and interactively.

In the chapters that follow, a variety of examples are considered. While tensions and challenges are identified, it is what made these examples work that we believe is most important, along with insights for applying such examples on a wider scale.

The Chapters of This Book

Setting the foundation for the remainder of the book, John Falk and Lynn Dierking link the formal and informal science learning landscapes through the lens of an ecosystem in Chapter "Viewing Science Learning Through an Ecosystem Lens: A story in Two Parts". They argue that a community's children and youth live and learn within a variety of settings, inclusive of their homes, schools, informal/free-choice learning organisations and institutions and workplaces. These settings are also shaped by a range of innovations and often mediated by digital media. All of these considerations mean that the boundaries of when, where, why, how and with whom people learn science are becoming increasingly blurred. The chapter goes on to explain why three critical processes (3C's) need to occur in order to maximise learning within the ecosystem:

* *co-ordination* of different learning resources and opportunities,
* *customisation* of activities and roles that optimise niches that are not present in the ecosystem, and
* *connection* of science education opportunities.

Using the case of the *Synergies* longitudinal research project, they provide insights into how resilient science learning ecosystems can be developed that overcome traditional barriers and enable people to continue to learn science daily and throughout their lives.

Erminia Pedretti and Ana Navas-Iannini explore the relationships between school and museum communities via a controversial exhibit, "Preventing youth pregnancy" in Chapter "Pregnant Pauses: Science Museums, Schools and a Controversial Exhibiton". The controversial nature of this exhibit in conjunction with the associated communication framework provides some novel insights. In a similar fashion to Falk and Dierking's "3C's"—coordination, customisation and connection in Chapter Viewing Science Learning Through an Ecosystem Lens: A story in Two Parts—Pedretti and Navas-Iannini identify the importance of building connections between formal and informal providers, and youth culture, to open up pathways for change.

In Chapter "Encounters with a Narwhal: Revitalising Science Education's Capacity to Affect and Be Affected", Steve Alsop provides a highly personal narrative of his encounters with a narwhal at the Royal Ontario Museum. Together with Justin Dillon, Steve provides a series of reflections that draw on this experience to consider teaching, learning, affective responses and offer an important perspective on science education—one that embraces the subjective, the embodied, the emotional and the relational dimensions based on experiences and awareness.

Moving from specific contexts to more general examples, Sue Stocklmayer takes us on a journey through science communication in Chapter "Communicating Science". She explores what sort of science students who have no intention of pursuing a science career need, contending that in such situations, science needs to be communicated to students and the general public in memorable and relevant ways. Non-scientists do not need to understand the finer details of climate science, genetic modification or nanotechnology, for example, but they do need to understand the potential impacts of developments in these areas and what information can be trusted.

In Chapter "Reinvigorating Primary School Science Through School-Community Partnerships", Kathy Smith, Angela Fitzgerald, Suzanne Deefholts, Sue Jackson, Nicole Sadler, Alan Smith and Simon Lindsay provide insights from three case studies for fostering successful school-community partnerships that focus on providing contextually rich opportunities for learning science. The cases emphasise the importance of teachers' voice, agency and ownership in driving change, and identify four conditions that are necessary for engaging in partnerships with the community:

- recognising a need for change,
- someone or something enabling change,
- seeking out the right partner(s), and
- applying the learning from the partnerships to promote further growth.

These four conditions again provide some similarities with the 3 C's identified by Falk and Dierking in Chapter "Viewing Science Learning Through an Ecosystem Lens: A story in Two Parts", and highlight the importance of a nuanced exploration of how such learning opportunities can be optimised.

Leonie Rennie, Grady Venville and John Wallace offer some provoking thoughts about the science learning opportunities provided through socioscientific issues (SSI) that potentially have personal connections, such as natural disasters. In Chapter "Natural Disasters as Unique Socioscientific Events: Curricular Responses to the New Zealand Earthquakes" they specifically look at the example of earthquakes and take us through a number of ways of looking at how education can and does respond to natural disasters. They contend that an interdisciplinary cross-curricular approach with links to the community is a necessary precursor to dealing effectively with any socioscientific issue.

Taking a slightly different approach to dealing with SSIs, Peter Fensham and Jasper Montana in Chapter "The Challenges and Opportunities for Embracing Complex Socio-scientific Issues as Important in Learning Science: The Murray-Darling River Basin as an Example" use the case of the Murray-Darling Basin in Australia to explore the challenges and opportunities that complex SSIs present to science education. Using the concept of boundary work, they discuss how challenges associated with curriculum intentions, assessment and teacher professional development will need to be addressed if decision making about SSIs is to genuinely become an embedded outcome of science education.

In Chapter "Outreach Education: Enhancing the Possibilities for Every Student to Learn Science", Debra Panizzon, Greg Lancaster and Deborah Corrigan take us through an innovative example of outreach education for students in the form of the National Virtual School for Emerging Science (NVSES). This project focuses on bridging the gap between traditional school science lessons and outreach by enabling school students, regardless of geographical location, to connect with like-minded students in a virtual classroom using an online platform. Participation in the virtual classroom was voluntary, provided access to experiences outside the classroom and to experts in their field, and clearly focused on learning as an important outcome of the experience. The example showcases the shifts required by teachers and students when they move from traditional to digital teaching structures, and the fact that outcomes are often as much about the product of the learning as the process.

In Chapter "Using a Digital Platform to Mediate Intentional and Incidental Science Learning", Cathy Buntting, Alister Jones and Bronwen Cowie extend on the learning opportunities provided by digital platforms. They consider intentional (when the learner sets out to learn something) and incidental (an unintentional or unplanned) learning when designing an on-line resource to support science learning—the New Zealand Science Learning Hub. They suggest that both of these types of learning tend to be situated, contextual and social, while at the same time can be mutually supported. Digital platforms offer opportunities to link these types of learning, and examination of such opportunities is important given the growing reliance on such platforms as the main sources of science-related information for Western societies. These platforms also challenge the traditional divide of "formal" and "informal" learning given their accessibility at any time, from anywhere and for multiple and different purposes.

The use of technology to impact on learning of multiple types is also a focus in Chapter ""Meet the Scientist": How Pre-service Teachers Constructed Knowledge and Identities" by Gillian Kidman and Karen Marangio, who explore extending the use of "meet the scientist" strategy to form a partnership between preservice physics teachers in Australia and a scientific research community in the US. This online partnership then provided a learning experience aimed at being as authentic as possible for Year 12 (17 year old) physics students in an Australian High School. Kidman and Marangio explore the multiple learning sites for the pre-service teachers, with a particular emphasis on the varied individual and collective learning afforded by these different sites.

In the final Chapter "Trial-and-Error, Googling and Talk: Engineering Students Taking Initiative Out of Class", Elaine Khoo and Bronwen Cowie use the notion of a learning ecology to scope the learning strategies and resources that undergraduate engineering students used to supplement their formal laboratory and lecture learning about a computer-aided design software package. Students engaged in conversation with peers, lecturers and workplace supervisors, trial-and-error in their own time, worked through course support materials, and used YouTube videos and dedicated online professional discussion forums. A capacity for self-initiated and self-directed learning is essential for today's engineering (and other) graduates, and the development of a learning ecology that blurs the formal–informal learning boundary is something that lecturers and universities need to consider.

In drawing this volume together, it is clear seeing learning as a formal–informal binary is no longer adequate or appropriate. In modern society, the traditional idea of formal learning occurring in formal education institutions such as schools, and informal learning occurring out of schools, does not support our students to develop the capabilities they will need to be successful in their futures. Indeed the personalisation, collaboration and informalisation of learning (often associated with the informal aspects of learning) will be at the core of learning in the future, as highlighted by Khoo and Cowie in Chapter "Trial-and-Error, Googling and Talk: Engineering Students Taking Initiative Out of Class". Nor are the boundaries between formal and informal learning easily distinguished. It is for this reason we have titled this book *Navigating the changing landscape of formal and informal science learning opportunities*. Given the lack of defined boundaries, the notion of an ever-changing landscape is important, requiring us to think about the learning opportunities across the entire landscape and not just those located within classroom walls. Navigating this landscape also requires us to broaden our ideas about learning science.

References

Anderson, U., Kelling, A., Pressley-Keogh, R., Bloomsmith, M., & Mapple, T. (2003). Enhancing the zoo visitor's experience by public animal training and oral interpretation at an otter exhibit. *Environment & Behaviour, 35*(6), 826–841.

Conway, W. G. (1969). Zoos: Their changing roles. *Science, 163*(3862), 48–52.

Crane, V., Nicholson, H., Chen, M., & Bitgood, S. (Eds.). (1994). *Informal science learning: What the research says about television, science museums, and community-based projects*. Dedham, MA: Research Communications Ltd..

Falk, J. H., & Dierking, L. D. (2012). *Museum experience revisited*. Walnut Creek, CA: Left Coast Press.

Falk, J. H., Dierking, L. D., Osborne, J., Wenger, M., Dawson, E., & Wong, B. (2015). Analyzing science education in the United Kingdom: Taking a system-wide approach. *Science Education, 99*, 145–173.

Viewing Science Learning Through an Ecosystem Lens: A Story in Two Parts

John H. Falk and Lynn D. Dierking

Abstract Studying the multidimensional, dynamic and complex qualities of a community-wide science education system must begin by creating an expanded definition of what constitutes a public science education system. A system-wide approach recognises that formal education entities (early childhood, elementary, secondary and post-secondary schools) are critical and necessary components to life-long, life-wide and life-deep science understanding and participation, but even collectively they represent only a small part, both physically and functionally, of the entire system. In a community-wide science education system the entire array of possible science education resources must be considered as equal contributors to public science education. Such community-wide science education system can be likened to a science learning ecosystem.

This chapter is comprised of two complementary parts. In the first we (Falk and Dierking) provide further background and theoretical foundation for an ecosystem approach to science education, discussing its implications for research, practice and assessment. In the second we (Dierking and Falk) describe an effort to study, and then use the data collected, to lead a participatory, community-wide effort to redesign a science learning ecosystem in a diverse, under-resourced community in Portland, Oregon, USA.

Keywords Informal science learning · Free-choice learning · Science learning ecosystems

Two decades ago, St. John and Perry (1993) proposed a reconceptualisation of education resources in a community, arguing that in addition to schools and universities, there are myriad learning resources – libraries, community-based organisations, museums, parks, print and broadcast media and the Internet – all components of a single, complex educational infrastructure (see also Falk & Dierking, 2002). More

J. H. Falk (✉) · L. D. Dierking
Oregon State University, Corvallis, OR, USA
e-mail: John.Falk@oregonstate.edu; dierkinl@science.oregonstate.edu

© Springer International Publishing AG, part of Springer Nature 2018
D. Corrigan et al. (eds.), *Navigating the Changing Landscape of Formal and Informal Science Learning Opportunities*,
https://doi.org/10.1007/978-3-319-89761-5_2

recently, this idea of an educational infrastructure has been reframed as an 'ecosystem' that offers a range of opportunities for learners of all ages to engage in life-long, life-wide and life-deep science learning (National Research Council, 2015; Traphagen & Traill, 2014). The ecosystem notion has been expanded to include not only material resources like institutions and organisations, but also social ones such as social networks, peers, educators (in school and out of school), friends and family (Tal & Dierking, 2014).

The implication of this ecosystem perspective is that today's children and youth live and learn within a variety of settings and configurations that include their homes, schools, informal/free-choice learning organisations and institutions, and workplace environments, all shaped by a continuous stream of emerging scientific and technological innovations and mediated by rapidly evolving digital media. Collectively, these resources form a complex community learning infrastructure. However, communities, and the complex learning infrastructure of intersecting educational entities they contain, are not mere "backdrops" for science learning—they are dynamic learning environments in which people engage, interact and make sense of the science they encounter in their daily lives (Barab & Kirshner, 2001). In great part, this ecosystem perspective is the result of increasing evidence that children and youth develop their understanding of science concepts in and out of school through an accumulation of experiences from different sources at different times, using a variety of community resources and networks (Falk & Dierking, 2010; National Research Council, 2009). The result is that the boundaries of when, where, why, how and with whom children and youth learn science are becoming increasingly blurred.

Despite the increasing evidence that children and youth in the twenty-first century pursue scientific interests and develop understandings across the day and throughout a lifetime, current approaches to analysing and supporting science education efforts have, by and large, remained mired in twentieth century models. Most educational research and development efforts are not planned or designed to consider the multidimensional, dynamic and complex qualities of a robust community-wide system. Instead, they focus on documenting the individual activities and outcomes of specific educational entities either in "formal" (e.g., school) or "informal" (e.g., science centers, digital media, after-school programmes), or occasionally as limited interactions between the two (e.g., school-run summer programmes or field trips to free-choice/informal settings). Typically these educational events occur over relatively brief time frames, sometimes as short as a few hours, and occasionally as long as a school semester. Only rarely do time frames extend across an entire year, and even more rarely, multiple years. While these more limited approaches have provided valuable insights into the contributions of individual educational entities to science learning, such focused efforts fail to account for the totality and extent of the contingent, lifelong and diverse learning opportunities that exist for children and adolescents within a robust community-wide system.

Studying the multidimensional, dynamic and complex qualities of a community-wide science education system must begin by creating an expanded definition of what constitutes a public science education system. A system-wide approach recognises that formal education entities (early childhood, elementary, secondary and post-secondary schools) are critical and necessary components to life-long, life-wide and life-deep science understanding and participation, but even collectively they represent only a small part, both physically and functionally, of the entire system. In a community-wide science education system the entire array of possible science education resources must be considered as equal contributors to public science education. This idea is beginning to gain traction, at least in the United States, as suggested by two high-profile national reports: *Identifying and supporting productive STEM programs in out-of-school settings* (National Research Council, 2015) and *How cross-sector collaborations are advancing STEM learning* (Traphagen & Traill, 2014). Both reports recommend taking an ecosystem-wide approach to science education, one that connects science learning experiences across the day and over a lifetime. As defined by the NRC report, a learning ecosystem is "the dynamic interaction among individual learners, diverse settings where learning occurs, and the community and culture in which they are embedded" (National Research Council, 2015, pp. 1–2).

This chapter is comprised of two complementary parts. In the first we (Falk and Dierking) provide further background and theoretical foundation for an ecosystem approach to science education, discussing its implications for research, practice and assessment. In the second we (Dierking and Falk) describe an effort to study, and then use the data collected, to lead a participatory, community-wide effort to redesign a science learning ecosystem in a diverse, under-resourced community in Portland, Oregon, USA.

Part 1: Envisioning a Resilient Pubic Science Education Ecosystem

John H. Falk and Lynn D. Dierking

What Is an Ecosystems Approach?

A wide range of disciplines have taken systems approaches to understanding the complex workings of complex assemblages, with the ecological sciences amongst the earliest of these efforts. Like many fields of biology, ecology—the study of the interactions of the collection of organisms living within a specific geographic area—began as a descriptive science, but more than a quarter of a century ago ecologists began applying increasingly sophisticated modeling strategies as a device for organising and better understanding the nature of systems and their community dynamics (Morin, 2011). Since most biological communities are extraordinarily complex, ecologists often focused their community investigations on conspicuous, readily identifiable sets of organisms, analysing the position they occupy in a food chain or other readily measurable relationship between them. This approach enabled ecologists to develop basic understandings of the structure and functioning of the assemblage of key organisms within a community—which organisms live there, what activities/roles they play and what the network of relationships between organisms look like (Morin, 2011). Relationships are investigated on a range of spatial and temporal scales, including the distribution, structure, abundance, equitability and interactions between coexisting groups (Morin, 2011).

Over recent decades ecologists have studied how structures and patterns of interaction within an ecosystem generate healthier, more robust systems. An interesting finding has been the growing appreciation that, independent of the type of ecological community studied, more complex, integrated and collaborative systems tend to be more productive and more robust (cf., Gunderson, Allen, & Holling, 2010; Oliver et al., 2015). Concomitant with this finding is that productive and robust communities tend to be more able to withstand perturbations, referred to as coherence. Robust systems are characterised as having numerous, often redundant and reinforcing feedback loops that feed information and resources back into the system. They also often have critical thresholds or tipping points, times at which the behaviour of the system changes rapidly due to relatively modest changes in external conditions.

Diversity in a healthy system is reflected by more than just the number of species or organisations present in a system (Biggs et al., 2012; Morin, 2011). Over time ecologists have come to appreciate that the critical criteria for determining the diversity of an ecosystem is not merely the presence of a large number of individual species but rather the presence of a large number of complex assemblages of species—diverse systems contain large numbers of functional, inter-articulating groups of organisms. In particular it is the organisation of the assemblages of species into diverse "niches" (i.e., roles and opportunities organisms or populations try to optimise to respond to the distribution of resources and competitors) that actually determines the diversity of a system (Gell-Mann, 1994). Thus an analysis of an ecosystem's health and resilience begins with studying the diversity of entities that comprise the community and the ways those entities interact and fulfill roles within it. Equally important is determining whether a given community has "empty niches"—roles, resources and opportunities not currently being fully utilised (cf., Gunderson et al., 2010; Levins, 1998).

Over the past 20 years, this kind of ecosystem approach also has been applied to human communities. Only recently have educational researchers begun to appreciate the potential of this perspective for understanding and facilitating the process of educational change. For example, researchers have constructed models to explore the effects of policies that promote school choice in large urban school districts in the U.S. (Maroulis et al., 2010), Lemke and Sabelli (2008) advocated taking a systems approach to the planning and design of educational interventions, and we utilised community ecology frameworks to analyse the workings of the science education infrastructure of the United Kingdom (Falk et al., 2015).

Although the application of an ecosystem lens to educational research is relatively new, the use of similar terms within the social sciences is not. This obviously creates the opportunity for considerable confusion. For example, a wide range of social science researchers have borrowed the concept of "ecologies" to frame their research, including Bronfenbrenner's ecological systems theory (1979) and, more recently, learning scientists talking about learning ecologies (e.g., Barron, 2006). Although we certainly share the conceptual spirit of these previous uses of the term ecologies—the idea that learning is a complex phenomenon that needs to be understood as occurring within the context of a range of sociocultural and physical contexts, multiple factors and players—how we use of the term "ecosystem" transcends these merely descriptive uses. Our approach attempts to not only describe but analyse the ways in which children and youths' science learning can be successfully supported and studied by utilising specific analytical frameworks and system-wide approaches developed within the field of ecological sciences (e.g., Falk, 1976; Morin, 2011; Oliver et al., 2015) and now applied to human systems (e.g., Falk et al., 2015; Folke et al., 2004; McAslan, 2010).

Defining a Resilient Science Education Ecosystem

Within an educational context, the goal of an ecosystem-based approach should be to create a public education system that meets the multiple science learning needs of a broad diversity of learners, supporting those needs 24 h a day, 365 days a year, year after year. In addition, the educational system, defined as the total community-wide set of educational offerings, needs to be able to support science learning year after year despite ever-changing economic, social and political conditions. For example, in recent years, school systems have been notoriously vulnerable to changing political and fiscal policies and priorities. At times science instruction is highly valued and prioritised, and at other times it has taken a backseat to "basics" such as reading and mathematics. In a robust educational system, young learners have science learning options regardless of the prevailing emphasis on science within schools. Ecologists call this type of robustness and redundancy *resilience*.

With recent interest in environmental sustainability there has been considerable focus among ecologists on the concept of resilience. Although resilience has at times been contested, with critics arguing it is an ambiguous, contradictory term that raises unresolved questions.

Just as in biological systems, the resilience of human communities has been found to be dependent on networks of interactions and synergistic actions based on systems of relationships, reciprocity, trust and social norms (Mahonge, 2010; McAslan, 2010). These networks and interactions are, in turn, influenced by the underlying diversity of the community (Holland, 2006). Because system-wide approaches to science education research have rarely been applied, most of the existing research affords researchers, practitioners and policy makers little or, at best, a superficial understanding of whether, and if so how, the various components of a robust, community-wide science education system might work collectively.

Based on research conducted in the United Kingdom, we (Falk et al., 2015) have previously postulated that viewed from the "top-down", a healthy and resilient science education ecosystem would be one in which the myriad science educational entities within the system are robustly inter-connected in a series of reciprocally beneficial relationships. The learners in such a system would be able to readily select from a vast smorgasbord of science learning offerings, with science education opportunities being available across multiple media platforms, for virtually any topic, presented at any level of depth or complexity and scaled to the unique developmental, intellectual, social, economic and cultural needs of any learner. The science education providers within the system would not only share overlapping goals, they would work synergistically towards achieving these shared goals. A characteristic of such a resilient system, analogous to a resilient natural ecosystem, would be a high diversity of educational providers and science education niches—a diversity of niches too great to be described by the overly simplistic dichotomy of formal and informal education providers. Finally, such a system, by virtue of its complexity and redundancy, would be able to withstand repeated perturbations, be they political or economic.

Meanwhile, from the "bottom-up", the learners in such a system would be able to pursue their interests and needs across the day, week, year and lifetime. They would be able to readily find the right educational platform—media, level of difficulty, and amount of personal mediation required—when and as they need it. They would be able to seamlessly move from one science learning experience to the next without perceiving that their experiences are fragmented or somehow discontinuous. Learners would therefore be able to pursue an uninterrupted and continuously reinforcing lifelong science learning journey; a journey that values each person's lived experience and builds on that lived experience in developmentally, culturally, intellectually and socially appropriate and relevant ways. Again, the current evidence is that this ideal is only possible now for the most affluent and educated citizens (Pew, 2015), and even for these individuals, considerable effort is required to fulfill this reality.

Based on ecological sciences research and our work in the UK, we propose that there are, at a minimum, three critical processes that need to be built into a resilient ecosystem of science education. These three processes can be summarised as an effort to: (1) *coordinate* different learning resources and opportunities within the ecosystem in such ways that learners are able to pursue their interests and needs across the day, week, year and lifetime; (2) *customise* activities and roles that optimise niches within the ecosystem, providing science learning offerings across multiple platforms for virtually any topic, presented at any level of depth or complexity and scaled to the unique developmental, intellectual, social, economic and cultural needs of any learner; and (3) *connect* science education opportunities through a series of reciprocally beneficial relationships that value diversity and redundancy in the system and the significant funds of knowledge that learners bring with them to each and every learning situation.

Coordinating Over the past decade or more, researchers have begun to develop strategies for understanding and documenting how learning develops, fluctuates and deepens across settings and over time (e.g., Ito et al., 2013; Stocklmayer, Rennie, & Gilbert, 2010). A growing number of studies demonstrate how individuals bring science understandings and practices developed in one setting to another (e.g., Azevedo, 2011; Barron, 2006; Falk & Needham, 2013). Thus, it makes sense to create a public education system that actively seeks to intellectually and functionally coordinate science learning experiences across sector boundaries, whether happening inside or outside of school, or across multiple settings and modalities. Collectively the goal needs to be to ensure that learning experiences occur seamlessly across different settings, times and contexts.

Customising A growing number of investigators (e.g., National Research Council, 2011; Tai, Qi Liu, Maltese, & Fan, 2006) have argued that rather than focusing on content, science education should focus on building a strong foundation of interest, facility and comfort with science ideas, practices and fields. If we accept this premise, the question arises: How and when should this focus begin? Obviously, it is never too early to begin supporting interest in science learning, however it has recently been shown that science interest during early adolescence, particularly between ages 10 and 14 years, is critical to long-term involvement in further science education and careers (Maltese & Tai, 2011; Tytler, Osborne, Williams, Tytler, & Cripps Clark, 2008).

Despite considerable efforts in recent years, evidence suggest that improvements in facilitating and sustaining science interest and participation, particularly among those in poor, under-resourced communities, impacts have been at best modest (Griffiths & Cahill, 2009; OECD, 2012; Osborne, Simon, & Collins, 2003). Children who have an interest in science at the middle school level are more likely to be motivated learners in science. They are also more likely to seek out challenge and difficulty, use effective learning strategies, and make use of feedback (Barron, 2006; Renninger & Riley, 2013). In addition, they are more likely to persist in tasks over time and to expend effort to master them, particularly when they experience feelings of enjoyment and value for the activities in which they are engaged (Linnenbrink & Pintrich, 2000; Wigfield & Eccles, 2000). Thus, at the heart of a robust science education ecosystem is a diverse and quality set of educational opportunities sufficient to build on and nurture the intrinsic curiosity and interest that all children have about science topics early in life, but that tends to wane as children become adolescents. The science education ecosystem, as a whole, needs to be capable of supporting customised experiences sufficient to satisfy the specific needs and interests of individual children and youth.

Connecting Most science education efforts fail to fully account for or value the significant funds of knowledge that learners bring with them to each and every learning situation, including schooling (Civil, 2016; Moll, Amanti, Neff, & González, 2005). Often, educational efforts vastly underestimate the significant differences that exist, developmentally, socially, economically and culturally within learner populations, and thus the consequences these differences can have on science learning. In addition, even well-intentioned educators can fall into the trap of being insensitive to sociocultural biases that they bring to their efforts—often unwittingly privileging dominant-cultural perspectives (cf., Fensham, 2007; Zeidler, 2016). Being mindful of sociocultural differences opens the door to greater access, social justice and inclusion. Science learning experiences should be designed in ways that directly connect to the social and cultural beliefs and practices of learners, their families and the diverse communities in which they live.

We therefore argue that, over the long-term, effective science education can best be achieved when entire communities work collaboratively to *coordinate* educational offerings between and across the entire learning ecosystem; *customise* experiences in ways that meet the unique intellectual needs and interests of learners; and *connect* learning experiences to the cultural and social realities of the community and its members. We call these the "3 c's" of a successful ecosystem-wide education. Such systems are not only more likely to be effective, they are also more likely to be resilient to stress and perturbation.

A range of science education entities have long engaged in aspects of these processes. Schools take children on field trips to science-rich settings such as natural areas or informal/free-choice science institutions like science centres, zoos or natural history museums. Universities offer internships for youth or supported summer science programmes. Community organisations such as scouting and, in the U.S. context, organisations such as 4-H, Girls inc., MESA and the Boys and Girls Club, attempt to include science programmes as part of their after-school offerings. Such programmes are often specifically tailored to the social and cultural realities of local youth. Very occasionally, schools and community-based organisations form partnerships in an effort to provide multi-institutional instruction around a specific topic. But these efforts are typically short-term, conceptualised and executed in isolation and uncoordinated across entire communities. As described by Traphagen and Traill (2014) and the National Research Council (2015), there are now exceptions to this reality with a small but growing number of communities attempting to put in place at least some aspects of an ecosystem-level effort. Described in Part 2 of this chapter are the efforts we have made in one U.S. community to create a more coherent, ecosystem-based approach to science education.

Part 2: Synergies: A Case Study of an Ecosystem Approach to Science Education

Lynn D. Dierking and John H. Falk

In this part we describe a research-practice partnership that we have been engaged with for over 5 years in a diverse, under-resourced community in the Pacific Northwest of the U.S. We are studying, and then using the data we collect to lead, a participatory, community-wide effort to redesign science learning from an ecosystem perspective. Our goal is to use these data to better coordinate, customise and connect the activities of science education providers in school and out of school, as well as empowering youth and their families to understand and better navigate the science education ecosystem. By doing so, we are striving to measurably improve access to, and use of, science learning resources in and out of school for this community's youth and their families.

Overview of Synergies

Beginning with a planning year in 2010, initially with support from the U.S.-Noyce and Lemelson Foundations, and most recently with support from the National Science Foundation, a research team based at Oregon State University initiated the *Synergies* research-practice partnership in the under-resourced Parkrose neighborhood of Portland, Oregon. The premise of *Synergies* is that if one better understands how, when, where, why and with whom children access and use science resources in and outside of school, it will be possible to use that information to create a more effective and synergistic community-wide educational system.

We selected Parkrose as a study site because it is a small, relatively self-contained community with many of the educational resources found in any city—schools, museums, after-school programmes, libraries and parks, as well as the socio-economic challenges found in urban communities. Although technically a "neighborhood" in Portland, the Parkrose community is unique in many ways. Historically Parkrose was not a part of the Portland metropolitan area and it continues to be served by a single, independent public school district. The Parkrose School District has four elementary schools that feed into a single middle school and then into one high school. In terms of informal science resources in the community, Portland has a number of quality informal science-related education institutions/organisations (e.g., science center, zoo and children's museum), but these resources have admission fees and require extensive travel to reach (the city's public transportation poorly serves the Parkrose neighborhood). Parkrose itself has a branch library and parks. In addition, at the beginning of the project, a small number of community-based organisations offered some form of science-related after-school and summer programming, although primarily for elementary-aged children.

Parkrose also is geographically bound, cut off from the rest of Portland on three of its four boundaries by two major freeways, the municipal airport and the Columbia River. The fourth boundary is not a physical border, but a socio-economic one—Parkrose is in northeastern Portland, the area of the city with the highest level of poverty, unemployment, access to drugs and crime. Although the Portland metro area is primarily white, Parkrose is a majority, minority community. According to U.S. Census statistics (U.S. Census, 2014), Parkrose residents fall within the following broad demographic categories: 38% White; 24% Latino/a; 18% Asian; 12% African American; 5% Native American; and 3% of Pacific Island origin. The majority of Parkrose residents are low income (e.g., 79% of children at the middle school are of sufficiently low income that they qualify for free or discounted meals).

But these statistics belie the true diversity of the community. Although nearly 40% of Parkrose residents are classified as "white non-Hispanic", over half of these residents are recent immigrants from Eastern Europe, and the "Asian" category is roughly equally represented by immigrants from Vietnam, Korea, Thailand and several parts of China. Similar diversity is also found within the "Latino/a" category. Over 50 languages are spoken in the school district of about 1200 children, and many children live in homes in which English is not the first language. Although a small community by many standards, Parkrose is large enough to mirror the majority of the complex social and economic dynamics of major urban areas, yet small enough from a research and practice perspective to be manageable in both scope and scale for a project of this nature.

When we began the *Synergies* project in 2010 there were relatively few opportunities for Parkrose youth to engage in science experiences, particularly after school, on weekends and over the summer. Other than sports, there were no local after-school programmes for middle school aged children—including in science or the arts—and very few for elementary school aged children. Although Portland itself is home to a wealth of informal STEM opportunities, including a world class science center, zoo, children's museum and a range of other out of school STEM programmes, these resources were functionally unavailable to the youth of Parkrose. The major impediment is geography. Portland is known for having great public transportation, but because of the unique social, political and economic history of Parkrose, most public transportation ends at the boundary of the community. In other words, in theory, one can access all of these great Portland science education resources from Parkrose by public transportation, but in practice this is true only if you can get to the boundaries of the community to begin the journey. Although the Oregon Museum of Science & Industry is a roughly 15–20 min car ride from the Parkrose community it is a nearly two-hour public bus ride from most of Parkrose.

Beyond its physical and socio-cultural characteristics, Parkrose is an ideal study site for one other critically important reason—the Parkrose community welcomed our research team from the start. We spent the first planning year identifying key educational partners, including Parkrose School District and other educational partners: Oregon MESA (Mathematics, Engineering, Science Achievement), Mt. Hood Community College, Portland 4-H Youth Development, Port of Portland (Portland Airport Authority), Girls, Inc. (a U.S. national STEM education provider), Metropolitan Family Services (regional social services organisation), Oregon Museum of Science & Industry (OMSI) and Metro (regional parks and recreation). All community members and partners have been uniformly open to change, excited about the prospect of being a community-based research laboratory and committed to improving the lives of their community's children.

In order to take a learner-centered perspective and build an empirical foundation for a community-wide system, over the first five years of the project we collected science interest and participation data from a single youth cohort—roughly 200 children, from age 10/11 years old (5th grade)—as well as developed in-depth case studies for a subsample of 15 youth and their families. We also convened community meetings with partners and initiated planning for community-wide "interventions". The goal of this work was for *Synergies* project staff to directly engage as many of the education players in the community, as well as parents, community leaders, and youth themselves in the redesign of the Parkrose science learning eco-system. The major product of these efforts was the *Parkrose Science Education Plan* which is being collaboratively implemented with our recently-awarded NSF grant. This new funding enables us to leverage the research findings from the on-going longitudinal study in support of a systematic and systemic design phase in which we will experiment with various interventions to determine whether we can influence the system, broadly writ. Meanwhile, basic data collection to monitor youth interest and participation in science will continue. Our research questions are:

- What is the nature of the science-related interests of 10–14 year old youth living in a single urban community and what factors seem to influence whether STEM interest increases, stays the same or diminishes over time?
- What science-related activities do 10–14 year old youth living in a single urban community participate in and what factors seem to influence whether participation in these activities increases, stays the same or diminishes over time?
- Are there significant differences in the science interest and participation profiles for youth as a function of socio-cultural background factors (gender, race/ethnic-ity, economic circumstances); support and encouragement by parents, teachers and peers; participation in out-of-school science activities; science understand-ing; perceptions of the value of science; perceptions of youths' self-efficacy in science; or parent/youth aspirations in science?
- Do science education interventions that are customised to take into account spe-cific science interest profiles appear promising for sustaining science interest and participation outcomes?
- In convening a wide variety of informal and formal educators to collaborate over time on developing an integrated, ecosystem approach to fostering youth science interest and participation, what challenges were encountered, what was per-ceived as working well, and what was perceived as not being effective?

Coordinating Science Learning Opportunities Across Out-of-School, School, Home and Other Settings

The *Synergies* project is committed to improving the quality of science learning in Parkrose, and we have argued that a key reason for the current challenges within science education is a failure to recognise that quality science learning is best sup-ported through a healthy and robust community-wide ecosystem. In most communi-ties, certainly in the U.S., there is a lack of coordination and cooperation between science education providers, in and out of school, resulting in a fragmented and inefficient collection of science education efforts and resources. Concomitantly, residents often do not fully appreciate the myriad resources for engaging with sci-ence that already exist within their community, nor completely understand the potential of these resources. Additionally, youth and families may lack social capital and/or agency for engaging with these resources more regularly and effectively.

Unfortunately, this is the case in many diverse, under-resourced communities with high numbers of immigrant families, such as in Parkrose.

In conceptualising *Synergies*, the goal was to determine to what extent we could design a research study that would provide both fundamental understandings about how diverse youth in an under-resourced community become interested and engaged in science (or not), across the settings, situations and time frames they traverse, and whether these data could be used to engage a community in rethinking and redesigning the education system, writ large.

We took a learner-centered approach, studying science learning through the lens of our single cohort of approximately 200 early adolescents for whom we have parental consent and assent forms as they moved across and through their own personal science ecosystems—not only physical resources such as school classrooms, after-school programmes, libraries, parks and museums, but also social networks of friends, siblings, family and teachers (in school and outside school) and digital resources youth engage with, including *Minecraft* and digital search engines such as *Google*, which vastly expand the science ecosystem's boundaries. We used two data collection strategies: (1) a primarily closed-ended questionnaire administered annually to every youth in the cohort; and (2) intensive, in-depth case study data collected roughly monthly with a subsample of youth from the cohort.

As of the writing of this chapter, we now have 4 years of quantitative and qualitative data. The quantitative data provides a detailed record of the year-by-year science-related interests and behaviours of the vast majority of youth in the cohort. We also have detailed qualitative data from the subsample of youth included in the case studies; highlights of their contribution to the study are described in a later section of this chapter. Collectively, these data form the empirical foundation for an understanding of the science learning ecologies of this single cohort of youth and provide a rough outline of the boundaries and nature of the science learning ecosystem within which youth in Parkrose currently interact. In the design phase we have just begun we will be adding other cohorts to study the development and implementation of interventions in a systematic manner, all built on the research foundations of the initial 5 years of the project. In particular, we will focus on the 6–7th grade time period, when the most significant changes in science interest and participation occurred.

As suggested above, *Synergies* has worked to leverage productive community partnerships within this neighborhood, coalescing into one community-wide system as many of the key formal and informal STEM education players in Parkrose as possible, as well as some outside the community who have the potential to participate. The goal is to help transform the Parkrose community into a place where youth can have firsthand experiences with science phenomena and materials and engage in sustained involvement with science practices, and where these practices will be supported by an entire ecosystem of opportunity and support. This was not the reality of Parkrose when we began the project in 2010.

All along, a key principle of *Synergies* has been the desire to directly engage formal and informal practitioners, as well as parents, community leaders and youth themselves, in the research process. Project staff spent the first planning year meeting individually and collectively with key partners to engage them in conversations about the project's goals and develop a shared theory of change for the community. As suggested by Connell, Kubisch, Schorr and Weiss (1999), this activity clarifies for all parties—researchers and practitioners alike—areas of consensus and differences in beliefs about key mechanisms for improving community outcomes. We felt this was critically important since we were mindful that it was highly possible that the ways we researchers are thinking about constructs of science interest and participation, and how we hypothesised these might develop (or wane) over time and across settings, were not likely to be the same as community partners, including educators. This process also transcends whether practitioners are familiar with current research on interest and participation. Most practitioners utilise planning tools (e.g., curricula, grant proposals) that deal with these constructs in a very linear fashion; rarely are constructs like science interest and participation conceptualised as complex, multi-dimensional variables. Thus a key part of these early discussions with partners and stakeholders revolved around building a shared understanding of what was meant by youth science interest and participation, particularly within a whole ecosystem context.

We also discussed the nature of the data to collect; this approach was taken before collecting a single bit of data. Thus, although we began with a set of basic research questions, these have been modified through an iterative process with community members. Since one of our key goals is to improve collaboration between and among all of the different youth-serving organisations/institutions within the science ecosystem, this input by practitioners into the empirical foundations and process is essential. We have facilitated theory- and model-building, collecting data on community theories and models, sharing data with participants, and the creation of strategies for improvement.

We also have engaged practitioners in a multi-step process resulting in the development of a comprehensive *Parkrose Science Education Plan*. To create this plan we assembled senior leadership from all 16 of our existing partners—28 individuals in total. The meetings were structured so that the *Synergies* team shared research findings to date, built consensus around strategies, and organised working groups to develop the education plan. Subsequently, smaller working groups created specific research-based plans around specific "challenges" highlighted by the research (e.g., engaging parents, leveraging peer interest, etc.). The resulting Plan was circulated amongst all partners for comments and edits; a final version was approved by all 16 participating organisations. We will begin to implement this plan in 2016 with the new funding we have received.

A key goal of the *Parkrose Science Education Plan* is to enhance the science learning infrastructure of Parkrose. To accomplish this we have taken a two-pronged strategy: (1) recruit and encourage existing Oregon and Portland informal education entities to more actively support science programming in the Parkrose community; and (2) with the aid of a Community Coordinator, supported by the *Synergies* project, help promote youth science programming in the community, as well as broker relationships between Parkrose-area formal and informal organisations.

To this end, the *Synergies* project has had great success in encouraging educational partners, many described earlier, currently not working in Parkrose to seriously consider bringing their assets to the community. We have also had success in encouraging partners who are working in Parkrose, but not at the middle school level, to seriously consider working with Parkrose Middle School level. Thus, over the first 5 years of the project we have been able to significantly build the informal STEM infrastructure of the community at minimal added cost. Most of these partners have existing funding, either operational monies or other grants, and the goal from the beginning was to leverage these resources so as to build a sustainable programme, wherever and whenever possible, utilising existing human and financial assets.

One positive outcome of the changes we have helped to broker in the community is that Parkrose School District administrators (the Superintendent, Director of School Improvement and Principal of the Middle School) have an increasing under-standing and enthusiasm for the role that they need to play in this effort. Although they have been supportive from the beginning, seeing the local commitment of oth-ers towards the community has galvanised their support and interest and they have, over time, become active partners.

The *Synergies* Community Coordinator has assumed responsibility for the day-to-day efforts of organising and coordinating project efforts within Parkrose. At this early stage of the endeavor, having an individual willing and able to provide this coordinating function is critical. She provides on-the-ground coordination with partners, parents and community "connectors", serving as the key interface with the schools and making connections between teachers and after-school providers with the goal of supporting youth's seamless learning, in and out of school.

Customising to Support of Youths' Interests, Experiences and Practices

It has become widely accepted that successful citizenship in the twenty-first century increasingly requires a foundation of interest, facility and comfort with science ideas, practices and fields (National Research Council, 2011). Some individuals will build on this foundation to pursue science academically and professionally, while others will pursue science-related hobbies and pursuits. All will require this foundation to make informed political, social and economic decisions. Accordingly, to ensure the strongest possible science literacy platform, it is essential to broaden and deepen access to and participation in quality science education for all young people, and especially young people from communities or social groups who his-torically have not been fully represented in science fields and/or science-related hobbies/pursuits (e.g., low income, minorities, females).

As suggested earlier in this chapter, adolescence is a critical time for fostering and promoting science interest and participation. The goal of *Synergies* is to improve an entire community's understanding of how science interests develop and how sci-ence education providers within a community can support that interest develop-ment. Our focus on interest and participation is predicated on an understanding that interest and knowledge are tightly inter-related, each developing over time through participation in science learning activities (cf., Renninger & Hidi, 2016).

From the start, the study has centered on how youth, themselves, define interest and participation in science-related topics. We developed the questionnaire for this aspect of the study through an iterative process in which we drew on a large body of research on interest development (particularly in science), existing instruments, research on youth participation in science, reviews by project advisors, several of whom are experts in the field of science interest development (e.g., Ann Renninger and Robert Tai), input from the initial group of youth researchers whom we hired explicitly for this purpose, and cognitive interviews with five 10/11 year old youth living outside the study area but comparable in background (for details on instrument construction and content see Falk et al., 2016). We also "piloted" the initial versions of the questionnaire with these youth researchers, asking them to critically assess the questions for both content and wording. Their feedback proved invaluable in helping to craft an instrument that was perceived as both relevant and comprehensible to our diverse sample.

In addition, an initial sample of 20 youth and their families were recruited to participate in on-going case study data collection; five families either left the study or moved out of the neighborhood so we have a final sample of 15 youth. These individuals were selected on the basis of gender, race/ethnicity, income and geography to be broadly representative of the broader population in Parkrose.

Although we had to limit the scope and scale of what constitutes science interest and participation in the questionnaire due to the practical necessities of reasonable length and time for administering, the more qualitative case studies are a vehicle for broadening the lens to include as much diversity in perspective and definition of science as possible. These data add richness and context to understandings about the varied repertoires of practice early adolescent learners engage in during a typical day, how and why children's STEM interests develop and change during this period of time, and which factors contribute to changes in STEM interest and engagement (e.g., family, friends, awareness, availability of and access to community resources, social capital, geography). These data also validate and enrich the survey, modeling and community efforts of the project.

For example, we have discovered that a number of factors directly contribute to youth science interest and participation. Key among these factors are parental and peer support, participation in out-of-school activities, and the self-perception that science content and practices are relevant to their lives. These findings form the evidence base on which the *Parkrose Science Education Plan* was created and will underlie the intervention strategies we will be experimenting with over the next 5 years. Also emerging from these data is that, not surprisingly, not all youth have the same levels and trajectories of interest and participation. In fact, most of the overall decline in science interest observed in the data, comparable to that of other national studies (cf., Osborne et al., 2003), is attributable to about a third of youth whose STEM interest and participation sharply declined between the 5th and 7th grades. However, the interest and participation of another third of youth has remained relatively unchanged, while nearly a third of youths' interest and participation has actually increased significantly. In other words, traditional efforts designed to deal with declining youth STEM interest and participation may be overgeneralising. The *Parkrose Science Education Plan* is designed to customise experiences in ways that support each of these groups differently—building on interest and participation where it exists, and fostering it where it is lacking.

Connecting to Family, Cultural Practices and Community

Successful citizenship in the twenty-first century will increasingly require not just knowledge but a foundation of interest, facility and comfort with science ideas, practices and fields (National Research Council, 2011). Many youth, particularly those from less advantaged circumstances, may not possess either the awareness or the social capital to successfully navigate the science learning ecosystem in ways that ensure comfort with science ideas and practices (Bodovski & Farkas, 2008; McCreedy & Dierking, 2013). Thus a key part of *Synergies* has been efforts to intentionally broker new and ongoing learning opportunities, and to support parents in particular in knowing which organisations and opportunities exist in the Parkrose community. These efforts range from building easily accessible databases of existing resources to explicitly directing individuals to internships, apprenticeships or making introductions to individuals and organisations. In all cases, brokering is designed to expand the personal networks of both parents and youth to help them navigate the broader science education ecosystem. Working with existing Oregon mentoring efforts, *Synergies* has attempted to facilitate the creation of an interactive science mentor database—one that can match youth with adults or even older youth who share specific interests. The goal is to facilitate the development of expertise and mastery, something both formal and informal science education institutions are notoriously bad at supporting for youth of this age. Children whose interests in science are developed and sustained are likely to become adults who pursue science interests at home and at work. However, to be effective, science efforts need to not only be "interesting" at the most superficial level, they must be socially meaningful and culturally relevant. Youth need to have agency in their learning. At a minimum they need be full partners in the learning process and, when possible, leaders.

Case study development has been critical to this aspect of the study. Data collection included an in-depth interview with youth every 6–8 weeks. Most of these were conducted in the youths' homes. Most interviews centered on a variety of activities in which we engaged the child and their family. For example, each family was given a digital camera to record family time, "days in the life" of the child, and science resources in the community. We also had youth create interest timelines in which they visually depicted how an interest had developed (or waned) over time and what the factors were that may have contributed. Youth made initial interest timelines in Year 1 when they were in 5th grade and these have been revisited, revised and discussed over time. We also worked with these youth to map their daily and weekly activities within Parkrose and the greater Portland metro area.

These data are revealing the "on the ground" realities of a subset of Parkrose families and how income, social capital and race/ethnicity influence youths' science interest and participation pathways, out of school and, in some cases, in school. For example, 2 of the 15 case study youth live only a mile apart from one another, yet their lives in the first summer we interviewed them about their photos of family time were very different. Their photos varied in terms of the activities they engaged in, the role their family played in seeking out and/or supporting their interests, and the influence of peers. We also have observed the role of income, social capital and race/ethnicity on youths' perceptions of their science interest(s)-dis-interest; where activities related to science take place, and if outside the home, how youth get there; and how and why their interests might have changed over time.

Case studies also have shown some of the interplay between out-of-school and in-school activity. One case study youth was very interested in mathematics in 5th grade, sharing his "love" for the topic during the first interview. When asked during a subsequent interview if he had a weekend when he could do whatever he wanted, he said he would visit a math website, sharing that he started getting on math sites when he was in 3rd grade, and by 5th grade he used them almost daily. Although he did not have strong parental support for his interest (his mother did not even know that math was his favorite subject), he independently sought out math websites and spent many hours after school solving math problems on his computer that were not for school, most often alone but sometimes with his cousin. However, by the end of 6th grade, his love of math was starting to wane and he was identified as being Math Dis-interested; this dis-interest persisted through 8th grade, though he remained interested in science, technology and engineering. Case study analysis indicated that there were several factors that could be implicated in the decline of his math interest over time. Perhaps the lack of family support ultimately made a difference, or the fact that none of his peers participated with him. Another factor, identified both through the survey and case study interviews, was the role of self-efficacy or self-concept in math. This youth indicated he gradually lost interest largely because school math was becoming more difficult, particularly fractions, which he said "hurt him".

Although in this young man's case, it was likely some combination of these factors, case study findings suggest that a number of out-of-school factors, including parental and peer support and participation in out-of-school activities, were significant in explaining how science interest may develop or decline over time. In addition, to understand the varied repertoires of practice early adolescent learners engage in during a typical day, it is critical to take into account issues of income, social capital, geography and race/ethnicity. Over the next 5 years, we plan to use these findings to inform the development of targeted intervention strategies that better support long-term youth interest and participation in science that we hope may lead to lifelong engagement in science-related pursuits.

It is also critical that these efforts connect to the community itself. A central tenet of the *Parkrose Science Education Plan* is that the most likely path to educational transformation is the application of a systems approach, which effectively harnesses a community's strengths and capacities by leveraging synergies between existing social, cultural, physical and technological resources. What this means in practice is focusing on systems-level topics as a vehicle for engaging youth in meaningful science experiences in their own community. For example, among the efforts currently being considered is a range of projects involving PDX, the international airport that abuts Parkrose (in fact the land for the airport was "cut out" of the original Parkrose footprint). Historically, despite being a major presence and employer of local residents, there has been virtually no interaction between PDX and the broader community. Yet PDX, like all international airports, is the focal point for a host of economic, social and environmental challenges that plague most urban areas, including illegal immigration, crime, economic effects of globalisation (e.g., reduction in jobs created by the consolidation of airlines), introduction of invasive plants and animals and air, water and noise pollution—all which impact Parkrose greatly because of their proximity. *Synergies* project staff are working in collaboration with colleagues at PDX, Parkrose schools and a range of informal education partners, including Schools Uniting Neighborhood Schools, 4-H and MESA, to build real-world issues and solutions into in-school and out-of-school science programming, with a key focus on careers. This will be one of the interventions we will test over the next 5 years.

One other important comment must be made as we close the discussion about connecting educational interventions (as well as our research) to the broader social-cultural-political realities of youth and their families. Over the 5 years we have been privileged to work in Parkrose, we have observed the conditions and stresses of life in a diverse and under-resourced community, particularly through the lens of our case study research. Over the last decade, many Parkrose families, like those in other urban communities, have experienced the loss of homes and jobs and are struggling to make ends meet on a day-to-day basis. There also has been a continuous flow of immigrants into and out of the community, many coming and leaving, not out of choice, but necessity. These circumstances present tremendous challenges to the families trying to make a life in Parkrose, and to the educational partners in the community, both in school and out of school, attempting to meet the needs of this diverse and under-resourced community. Our research is influenced by these challenges too.

Conclusions

There is a revolution afoot! We are witnessing a tectonic shift in how, when, where, with whom and even why people learn science. Just as the information revolution dramatically transformed societies globally, this learning revolution is changing the way people live and learn in the twenty-first century. Science learning today is 24/7, continuous and on-demand. Whether aged 5 or 95, learners seek educational experiences from a myriad of sources, while at home, on weekends and even while on vacation. For the past 100 years as societies we have come to believe that the words "learning", "education" and "school" are synonymous—but today public science education does not just happen at school. Learners spend only a fraction of their lives in a classroom. In fact, research indicates that traditional achievement gaps are less a factor of disparities in classroom learning than inequities in access to enriching experiences in the out-of-school time space (Pew, 2016). Most learning is *free-choice*, driven by an individual's needs, interests and access to learning opportunities.

If we are to achieve the oft-stated goals of creating a science engaged and literate public than we need to invest in public science education—not exclusively a school-based public education system but one that occurs year-round, from morning to night and across the lifespan. We need to invest in creating a network of public science education experiences that seamlessly incorporate learning opportunities in and out of school, framed increasingly around opportunities that support each individual learner's desire to answer important questions in his or her own life. To do this, we need to build locally, nationally and increasingly globally robust and resilient ecosystems for science learning. Within such an ecosystem, learners will be able to coordinate their learning experiences across settings, times and topics; they will be able to leverage and customise community resources and partnerships to meet their own science learning goals; and they will be able to broker science learning opportunities across their lives in ways that connect to their family and community cultural practices and realities. The *Synergies* project in Portland, Oregon represents one example of how an entire community, with support from a major university, is working together in an effort to build such a science learning ecosystem. In the future, we must work towards the creation of many such public science education ecosystems— educational systems that move away from one-size-fits-all instruction to more individualised, competency-based experiences that encourage the exploration of new ideas while at the same time providing just-in-time, personalised support for individually relevant science learning for all.

References

Azevedo, F. S. (2011). Lines of practice: A practice-centered theory of interest relationships. *Cognition and Instruction, 29*(2), 147–184.

Barab, S. A., & Kirshner, D. (2001). Rethinking methodology in the learning sciences. *Journal of the Learning Sciences, 10*(1&2), 5–15.

Barron, B. (2006). Interest and self-sustained learning as catalysts of development: A learning ecology perspective. *Human Development, 49*, 193–224.

Biggs, R., Schlüter, M., Biggs, D., Bohensky, E. L., BurnSilver, S., Cundill, C., … West, P. C. (2012). Toward principles for enhancing the resilience of ecosystem services. *Annual Review of Environment and Resources, 37*, 421–448.

Bodovski, K., & Farkas, G. (2008). "Concerted cultivation" and unequal achievement in elementary school. *Social Science Research, 37*(3), 903–919.

Bronfenbrenner, U. (1979). *The ecology of human eevelopment: Experiments by nature and design*. Cambridge, MA: Harvard University.

Civil, M. (2016). STEM learning research through a funds of knowledge lens. *Cultural Studies of Science Education, 11*(1), 41–59.

Connell, J., Kubisch, A., Schorr, L., & Weiss, C. (1999). *New approaches to evaluating community initiatives* (Vol. 1). Washington, DC: Aspen Institute.

Falk, J. H. (1976). Energetics of a suburban lawn ecosystem. *Ecology, 57*(1), 141–150.

Falk, J. H., & Dierking, L. D. (2002). *Lessons without limit: How free-choice learning is transforming education*. Lanham, MD: AltaMira.

Falk, J. H., & Dierking, L. D. (2010). The 95% solution: School is not where most Americans learn most of their science. *American Scientist, 98*, 486–493.

Falk, J. H., Dierking, L. D., Osborne, J., Wenger, M., Dawson, E., & Wong, B. (2015). Analyzing science education in the U.K.: Taking a system-wide approach. *Science Education, 99*(1), 145–173.

Falk, J. H., Dierking, L. D., Staus, N. L., Wyld, J. N., Bailey, D. L., & Penuel, W. R. (2016). The Synergies research–practice partnership project: A 2020 Vision case study. *Cultural Studies of Science Education, 11*(1), 195–212.

Falk, J. H., & Needham, M. D. (2013). Factors contributing to adult knowledge of science and technology. *Journal of Research in Science Teaching, 50*(4), 431–452.

Fensham, P. J. (2007). Values in the measurement of students' science achievement in TIMSS and PISA. In D. Corrigan, J. Dillon, & R. Gunstone (Eds.), *The re-emergence of values in science education* (pp. 215–229). Rotterdam, The Netherlands: Sense.

Folke, C., Carpenter, S., Walker, B., Scheffer, M., Elmqvist, T., Gunderson, L., … Holling, C. S. (2004). Regime shifts, resilience, and biodiversity in ecosystem management. *Annual Review of Ecology, Evolution and Systematics, 35*, 557–581.

Gell-Mann, M. (1994). *The quark and the jaguar: Adventures in the simple and the complex*. San Francisco, CA: W.H. Freeman.

Griffiths, P., & Cahill, M. (2009). *The opportunity equation: Transforming mathematics and science education for citizenship and the global economy*. New York, NJ: Carnegie Corporation of New York and Institute for Advanced Study.

Gunderson, L., Allen, H. R., & Holling, C. S. (Eds.). (2010). *Foundations of ecological resilience*. Washington, DC: Island.

Holland, J. E. (2006). Studying complex adaptive systems. *Journal of Systems Science & Complexity, 19*(1), 1–8.

Ito, M., Baumer, S., Bittanti, M., Boyd, D., Cody, R., Herr-Stephenson, B., … Tripp, L. (2013). *Hanging out, messing around, and geeking out: Kids living and learning with new media*. Cambridge, MA: MIT.

Lemke, J. L., & Sabelli, N. H. (2008). Complex systems and educational change: Towards a new research agenda. *Educational Philosophy and Theory, 40*(1), 118–129.

Levins, S. A. (1998). Ecosystems and the biosphere as complex adaptive systems. *Ecosystems, 1*, 431–436.

Linnenbrink, E., & Pintrich, P. R. (2000). Multiple pathways to learning and achievement: The role of goal orientation in fostering adaptive motivation, affect, and cognition. In C. Sansone & J. Harackiewicz (Eds.), *Intrinsic and extrinsic motivation: The search for optimal motivation and performance* (pp. 195–227). San Diego, CA: Academic.

Mahonge, C. (2010). *Co-managing complex social-ecological systems in Tanzania. The case of Lake Jipe wetland* (e-book). Wageningen, The Netherlands: Wageningen Academic.

Maltese, A. V., & Tai, R. H. (2011). Pipeline persistence: Examining the association of educational experiences with earned degrees in STEM among U.S. students. *Science Education, 95*(5), 877–907.

Maroulis, S. J., Guimerà, R., Petry, H., Stringer, M. J., Gomez, L. M., Amaral, L. A. N., … Wilensky, U. (2010). Complex systems view of educational policy research. *Science, 330*(6000), 38–39.

McAslan, A. R. R. (2010). *The concept of resilience: Understanding its origins, meaning and utility*. Adelaide, Australia: Torrens Resilience Institute.

McCreedy, D., & Dierking, L. D. (2013). *Cascading influences: Long-term impacts of informal STEM experiences for girls*. Philadelphia, PA: The Franklin Institute.

Moll, L. C., Amanti, C., Neff, D., & González, N. (2005). Funds of knowledge for teaching: Using a qualitative approach to connect homes and classrooms. In N. González, L. Moll, & C. Amanti (Eds.), *Funds of knowledge: Theorizing practice in households, communities, and classrooms* (pp. 71–87). Mahwah, NJ: Lawrence Erlbaum.

Morin, P. J. (2011). *Community ecology* (3rd ed.). New York, NY: Wiley-Blackwell.

National Research Council. (2009). *Learning science in informal environments: Places, people and pursuits*. Washington, DC: National Academy.

National Research Council. (2011). *A framework for K-12 science education*. Washington, DC: National Academy.

National Research Council. (2015). *Identifying and supporting productive STEM programs in out-of-school settings*. Washington, DC: National Academy.

Oliver, T. H., Heard, M., Isaac, N., Roy, D., Procter, D., Eigenbrod, F., … Bullock, J. M. (2015). Biodiversity and resilience of ecosystem functions. *Trends in Ecology & Evolution, 20*(11), 673–684.

Organization for Economic Co-operation and Development (OECD). (2012). *PISA in Focus 18: Are students more engaged when schools offer extracurricular activities?* Paris, France: OECD.

Osborne, J., Simon, S., & Collins, S. (2003). Attitudes toward science: A review of the literature and its implications. *International Journal of Science Education, 25*(9), 1049–1079.

Pew. (2015). *The American middle class is losing ground. Technical report*. Retrieved from http://www.pewsocialtrends.org/2015/12/09/the-american-middle-class-is-losing-ground/

Pew (2016). *State of the News Media 2016*. http://assets.pewresearch.org/wpcontent/uploads/sites/13/2016/06/30143308/state-of-the-news-media-report-2016-final.pdf   Retrieved December 18, 2016.

Renninger, K. A., & Hidi, S. (2016). *The power of interest for motivation and engagement*. New York, NY: Routledge.

Renninger, K. A., & Riley, K. R. (2013). Interest, cognition, and the case of L__ in science. In S. Kreitler (Ed.), *Cognition and motivation: Forging an interdisciplinary perspective* (pp. 325–382). New York, NY: Cambridge University.

St. John, M., & Perry, D. (1993). A framework for evaluation and research: Science, infrastructure and relationships. In S. Bicknell & G. Farmelo (Eds.), *Museum visitor studies in the 90s* (pp. 59–66). London, UK: Science Museum.

Stocklmayer, S. M., Rennie, L. J., & Gilbert, J. K. (2010). The roles of the formal and informal sectors in the provision of effective science education. *Studies in Science Education, 46*(1), 1–44.

U.S. Census Bureau. (2014). *Current population survey, 2013 annual social and economic supplement*. Washington, DC: Department of Labor.

Tai, R. H., Qi Liu, C., Maltese, A. V., & Fan, X. (2006). Planning early for careers in science. *Science, 312*, 1143–1145.

Tal, T., & Dierking, L. D. (2014). Editorial: Learning science in everyday life. *Journal for Research in Science Teaching, 51*(3), 251–259.

Traphagen, K., & Traill, S. (2014). *How cross-sector collaborations are advancing STEM learning*. Los Altos, CA: Noyce Foundation.

Tytler, R., Osborne, J., Williams, G., Tytler, K., & Cripps Clark, J. (2008). *Opening up pathways: Engagement in STEM across the primary-secondary school transition*. A report commissed by the Australian Department of Education, Employment and Workplace Relations. Retrieved from https://docs.education.gov.au/system/files/doc/other/openpathinscitechmathenginprim-secschtrans.pdf

Wigfield, A., & Eccles, J. S. (2000). Expectancy-value theory of achievement motivation. *Contemporary Educational Psychology, 25*, 68–81.

Zeidler, D. (2016). STEM education: A deficit framework for the 21st century? A sociocultural response. *Cultural Studies of Science Education, 11*(1), 11–26.

Pregnant Pauses: Science Museums, Schools and a Controversial Exhibition

Erminia Pedretti and Ana Maria Navas-Iannini

Abstract Recently, there have been movements towards the inclusion of critical and often controversial exhibitions in science centres and museums. In this case study we consider the controversial exhibition *Preventing Youth Pregnancy,* hosted by the Catavento museum (São Paulo, Brazil). Specifically, we explore responses from, and relationships between, school and museum communities that attended the exhibit. We begin with a brief literature review on informal settings and controversial exhibitions, and present a science communication framework that informed our research. Findings are framed by three major themes: *building connections between the formal and the informal sector through collaboration, building connections with youth culture,* and *building pathways for change.* The chapter concludes with a discussion about the challenges faced by museums and science centres in creating and/or displaying controversial exhibitions.

Keywords Controversial exhibitions · Science museums · Youth pregnancy · Sexuality · Socioscientific issues · Brazil

Introduction

In this chapter we explore responses from, and relationships between, school and museum communities, in the context of the controversial exhibit *Preventing Youth Pregnancy* displayed at the Catavento Cultural and Educational museum in São Paulo, Brazil. This work is part of a larger funded project entitled *Engaging the Public with Controversial Exhibitions at Science Centres and Museums.* The overall purposes of this project are to: undertake a critical analysis of exhibitions that are controversial in nature or issues-based; explore the interface between science communication and visitor engagement (visitors being adults, children, school groups); and examine theoretical and practical considerations for museums and science centres as they design and/or host controversial exhibitions. A series of case studies

E. Pedretti (✉) · A. M. Navas-Iannini
Ontario Institute for Studies in Education, University of Toronto, Toronto, Canada
e-mail: ana.navasiannini@mail.utoronto.ca; erminia.pedretti@utoronto.ca

© Springer International Publishing AG, part of Springer Nature 2018
D. Corrigan et al. (eds.), *Navigating the Changing Landscape of Formal and Informal Science Learning Opportunities,*
https://doi.org/10.1007/978-3-319-89761-5_3

are underway at institutions across Canada, and internationally; this particular piece represents one of the case studies we have developed to date. Through activities that make use of drama and narrative, the exhibit *Preventing Youth Pregnancy* addresses highly critical socioscientific issues such as sexual practices, prevention of sexually transmitted diseases, and unexpected pregnancies. In this research we focus on visiting teachers' and students' experiences with the exhibit, as well as curators' and museum educators' expectations about it.

Relationships between science museums and schools are well documented (see, for example, Griffin, 2004; Pedretti, 2004; Stocklmayer, Rennie, & Gilbert, 2010). However, less is known about the interactions between such communities and *controversial* exhibitions (for examples of research in this area, see Barrett & Sutter, 2006; Macdonald, 1998; Pedretti, 2002). Accordingly, we begin the chapter with a brief review of the recent movement towards the inclusion of critical (and often controversial) exhibitions in science centres and museums, and discuss a science communication framework that informed our research. This is followed by a description of our case study, the exhibition *Preventing Youth Pregnancy*. In the final sections we present and discuss our findings, which are framed by three major themes: building connections between the formal and the informal sector through collaboration, building connections with youth culture, and building pathways for change. We conclude the chapter with an examination of the challenges inherent in displaying controversial exhibitions, and how museums and science centres might respond to these challenges.

Critical Exhibitions: Trends in Museum Practices

Historically, science museums have focused on exhibitions that represent science as objective, unproblematic, without context, separated into disciplines, and supported by a top-down model of knowledge transmission (Bradburne, 1998; Delicado, 2009; Janousek, 2000). Typically, they emphasise cultural heritage through artifacts, collections, object displays, and curiousity cabinets—extoling the wonders and virtues of science to the public (Pedretti & Dubek, 2015). However, in recent years, informal settings like science centres have witnessed increased attention to issues in science and technology, and consequently have moved in new and bold directions. Some have begun to develop contemporary and often provocative installations (e.g., *Body Worlds*; *A Question of Truth*; *Renewable Energies: Time to Decide*; *Sex: A Tell All*) with all the social and political trappings of the day. This has led to the emergence of a category of installations described as *critical exhibitions* (Pedretti, 2002, 2004, 2012; Pedretti & Dubek, 2015). Critical exhibitions are typically issues-based (e.g., genetically modified foods, renewable energy, climate change, reproductive technologies) and embedded in rich social, cultural and political contexts, often exposing and/or generating controversy. Visitors are invited to consider socioscientific materials from different perspectives and multiple points of view, and to grapple with their own positionality. Such exhibitions also pose interesting questions about the ways in which science is (re)presented to the public (Hodder, 2010; Macdonald & Silverstone, 1992). These exhibits frequently make use of narratives, role-play, simulations, dramatisation, and fictional stories, and create opportunities of 'being in-the-place-of' as ways of engaging diverse audiences with complex subject matter, while also learning to care about others. Hodson (2014) eloquently describes how stories and dramatisation can be powerful learning opportunities. We extrapolate his sentiments to the informal context:

> Through stories, and especially through drama, students are stimulated to address issues and events from the perspectives of others, explore and develop understanding, establish new relationships and consolidate existing ones. In other words, engaging with narrative is as much a way of knowing ourselves as it is a way of understanding the views of others. (p. 78)

Critical exhibits can easily reach a controversial locus due to the complex and emotionally and politically charged nature of their subject matter, and due to the variety of responses and points of view that they generate in visitors and communities. In spite of their complexity, critical and controversial displays can serve many purposes. They provide a useful context for: (1) contesting the status quo and the ways in which science is constructed by including, for example, debates and controversies within the scientific community; (2) approaching socioscientific issues from a variety of perspectives; (3) raising awareness of the political, economic and environmental forces in which science is embedded; (4) inviting visitors to explore the nature of science and the relationships among science, technology, society and environment; (5) challenging visitors' points of view; (6) involving struggles over meaning and morality, power and control; (7) stimulating visitors to actively participate and engage; (8) creating spaces for visitors to be heard; (9) bringing visitors closer to contemporary science; and (10) constructing more equality relations between exhibitors and visitors (Delicado, 2009; Durant, 2004; Einsiedel & Einsiedel, 2004; Mazda, 2004; Pedretti, 2002 2004; Pedretti & Dubek, 2015).

Furthermore, critical and controversial exhibitions have the potential to shift from passive to more iterative communication approaches that allow for consultation, knowledge co-production, and social responsibility (Bucchi, 2008; Gascoigne, Cheng, Claessens, Metcalfe, & Schiele, 2010; Lewenstein, 2003; Trench & Bucchi, 2010). The field of *science communication* describes these latter interactions as being a central part of the dialogic and participatory models of communication (Bucchi, 2008), which present powerful learning opportunities for engaging museum visitors. In this study, the emergent field of *science communication* provides a useful lens for deconstructing representations of science, while shedding light on how visitors interact and engage with particular installations. In the next section, we discuss science communication models in the context of science museums.

Science Communication and Museum Exhibits

> It may be expected, for instance, that an issue in the field of particle physics, with low pub-
> lic impact and mobilisation, little controversy among experts, propelled by visible research
> institutions ... may lend itself to a deficit-like pattern in which the public is invited and
> willing to appreciate the spectacle of science's achievements. Likewise, an issue such as
> genetically modified organisms, touching many publicly relevant themes including food,
> safety, biodiversity and resource distribution, with a certain amount of experts' disagree-
> ment as publicly perceived, propelled by corporate actors in a context highly sensitive,
> alerted and mobilised to questions of environment and globalisation, was [sic] unlikely to
> be containable in the deficit box. (Bucchi, 2008, p. 71)

Massimiano Bucchi's words raise key issues regarding the positioning of controversial topics within the field of science communication. According to Bucchi, stable scientific topics can more easily lead to passive communication approaches
("the deficit box"), while complex and controversial topics are more likely to allow
other forms of communication to emerge. The deficit model, characterised by a top-
down communication approach that flows from the specialists to the non-specialists,
emerges from the notion that experts need to fill the knowledge gap in the lay public
while simultaneously creating (in the views of its supporters) a more welcoming
and positive climate to scientific development. Critics of the deficit model advocate
for more iterative approaches to science communication that could offer different
ways of engaging society with current and critical scientific issues (House of Lords,
2000). More recent communication models emphasise two-way communication
approaches, and focus on, for example, implications of research and social responsibility, and agency (for more, see chapter "Communicating Science" by Sue
Stocklmayer).

By way of summary, Bucchi's (2008) model of science communication[1] includes
three dimensions: *deficit, dialogue* and *participation*. Moving beyond the idea of
transferring scientific content and knowledge in a one-way mode (often deficit), iterative models of science communication emphasise dialogue, consultation, negotiation,
knowledge co-production and participation. These latter ideas echo the position of
the American Association for the Advancement of Science (AAAS, 2014) regarding
calls for *public engagement* and the creation of opportunities for public dialogue.

In a parallel movement, there have been significant calls for a science education
that includes ideas of participation, democratic citizenship and action (e.g., Bencze
& Alsop, 2014; Hodson, 2014; Levinson, 2010). These ideas have been helpful in
shaping our work with controversial issues in science and technology in the public
domain. For example, Levinson's (2010) framework for democratic participation
includes the following dimensions: *deficit, deliberative, science education as praxis,*
and *dissent and conflict*. While Bucchi's (2008) dialogue and participation
dimensions emphasise *context and content*, Levinson's notion of *dissent and con-*

---

[1] The *models of science communication* represent one of the most important theoretical contributions in the emergent field of science communication (Gascoigne et al. 2010; Trench & Bucchi,
2010) and they have been used to describe and analyse the ways in which science and society
interact. For more information, descriptions and discussions about those models, see Durant
(2004), Lewenstein (2003), Pouliot (2009), Schiele (2008), and Trench and Bucchi (2010).

Table 1 Merged dimensions of science communication

Dimensions of science communication	Emphasis	Communication model	Aims
Deficit	Content	Transfer, popularisation, one-way, one-time	Transferring knowledge
Dialogue	Context	Consultation, negotiation, two-way, iterative	Discussing implications of research
Participation	Content and context	Knowledge co-production, deviation, multi-directional, open-ended	Setting the aims, shaping the agenda of research
Dissent and conflict	Content and context (political literacy)	Knowledge is distributed, emergent on a need-to-know-basis, multidirectional, open-ended	Political understanding and action for changing the research agenda

flict relates to political literacy through action. Recognising the similarities between Bucchi's and Levinson's works, we merged the two (see Table 1) in an effort to help us understand the relationship between science communication models and democratic participation in science education, in the context of controversial exhibitions and the visitor experience.

The dimensions of science communication presented in Table 1 also relate to the *engagement continuum* for science museums activities developed by Einsiedel and Einsiedel (2004). In their continuum, visitors' involvement and engagement can range from passive to (inter)active. The *passive* extreme represents more traditional roles of museums, where scientific narratives are presented without a context and where visitors are invited to observe, contemplate and "receive" information. The *interactive* pole reflects ideas of science in context, social empowerment and responsibility, and supports a more iterative and participatory level of visitor engagement. Moreover, visitors' personal contexts—experiences, perceptions, beliefs, concerns and values—are important, as they begin to express their positions, critically discuss the issues presented and make decisions about them (see Bell, 2008). Critical exhibitions are located at the interactive pole of this continuum in that they provide spaces that allow visitors to join debates and express opinions about the issues (often controversial) at hand (Mazda, 2004). Supported by these considerations, we argue that science museums can play an important role in fostering more iterative and participatory levels of visitor engagement through exhibitions that support critical thinking, dialogue, current research and complexity.

"Preventing Youth Pregnancy": A Provocative and Controversial Exhibition

Preventing Youth Pregnancy is an exhibition that delves into the issue of teenage pregnancy, associated risks and prevention. The display is part of São Paulo's Catavento Cultural and Educational museum's second floor area called *Society*, where topics that cut across science, technology, society and environment are presented to the public (e.g., nanotechnologies, science and politics, drug consumption). This exhibit was developed during 2008–2009 through a partnership between the Kaplan Institute,[2] the State Secretary of Education and the museum. The exhibit has been on display since 2009 and has experienced some modifications. However, its goals have remained constant, as one of the curators describes:

> I believe that the goal is the same, the goal is to sensitise [youth] about the impact of a pregnancy and a STD [sexually-transmitted disease], (which is)…forever. I believe that this is extemporary. (Interview, female, 30s, member of the curatorial team of the exhibit)

Located in a bright and somewhat hidden room of the museum's old building, the exhibit begins with an empty space, equipped simply with purple puff pillows in the middle of the room (Fig. 1a). Visitors, older than 13 years of age, are invited to sit on the pillows and the museum sex educator begins with some welcoming remarks and a brief introduction to the space. He/she then initiates a conversation that focuses on visitors' plans and hopes for their future. This is accomplished through a few warm up activities that include: a *sharing* moment (visitors are invited to share why they chose to come to that exhibit); a *projection* moment (visitors are taken on a trip to the future through the use of audio-visual resources); and, finally, an *introspection* moment (visitors are provided with some quiet time to reflect on, and write down on paper, their dreams for the future; they carry this paper with them throughout their entire visit). The intention is to elicit from students what they

Fig. 1 (a) This is how the exhibit first looks to visitors as they enter, and (b) how it is then transformed into a party/labyrinth

---

[2] The Kaplan Institute created in 1991 is a Brazilian institute for studies in human sexuality. Its aims are: therapeutic treatments for sexual difficulties, sexual education, health education and sexual responsibility. In 2006, the institute focussed exclusively on sexual education, particularly sexual education for teenagers (http://www.kaplan.org.br/).

Fig. 2 Image of an exhibit panel. The translation of the text posed to visitors reads as follows (at the top of the panel): "If you came from panel 02: "Attention! If you put others at risk your vulnerability for a STD increases too". (On the bottom of the panel and related to the new fictional situation): "Today is the big day: you two are going to have sex. You have the most important thing—the condom, which has been in your bag for a while: (*A*) You use the condom you have in your bag, (*B*) You use a new condom"

envision for their future, where they see themselves after finishing school and how they expect their lives will unfold. At this point, the sex educator invites them to a party. The same room, previously bright and spacious, becomes dark, music is turned on and the exhibit emerges through walls to form a labyrinth (Fig. 1b). Each wall or panel presents a situation that involves sexual practices in a frank and direct manner and invites visitors to make decisions (Fig. 2, see also Appendix).

According to the choices visitors make, they move through different paths in the labyrinth to engage in new fictional scenarios. The text on this new panel is still related to the visitor's choice (e.g., if the visitor chose to use a new condom instead of the old one she has carried for long time in her bag, the message will emphasise that with this choice unexpected pregnancies may be avoided). While they move (on their own) and enjoy 'the party', scenarios appear, such as STDs or pregnancies A museum educator describes the experience as follows:

> And then, a party starts from scratch. And, so, we turn on the lights, we play music for the party and then we pull a labyrinth from the ceiling and, then, they [the visitors] answer the questions… they go through 18 situations that represent risk and, based on their responses, they are directed to other panels. If they go to panels 10 or 11 they might be infected with AIDS… a STD or they might get pregnant and so, when they get pregnant they have to wear a balloon under the t-shirt. When the party is over and the panels go up to the ceiling again, you see their reaction. (Interview, male, 18–20s, museum educator)

When the party is over, the room is transformed again into a meeting space and visitors' plans and hopes for the future are re-visited in a final conversation/forum facilitated by the sex educator. During this final meeting, there is also an opportunity for students (and visitors in general) to ask questions, resolve doubts and fill any knowledge gaps related to topics raised by or in the exhibit.

Research Background

We adopted a naturalistic case study methodology (Lincoln & Guba, 1985) using qualitative methods of data collection, such as interviews and observation, which are commonly used in museum research (Diamond, Luke, & Uttal, 2009; Soren & Armstrong, 2014). We gathered data that would allow us to produce rich, thick descriptions of the exhibit, visitor responses and museum staff perceptions. Data for the exhibit *Preventing Youth Pregnancy* were collected between November 2014 and February 2015 and included open-ended interviews with museum staff and visiting publics, observations of visitors interactions with the exhibit, field notes, and collection of documents.

In this chapter we focus on data that represent the voices of museum staff, teachers and students. These data included (i) five interviews with museum visitors, two teachers and three students; (ii) observations of and field notes about the interviewees; (iii) five interviews with museum staff (educators who work in the exhibit), and a member of the curatorial team of the exhibit; and (iv) 40 exit comment cards completed by school teachers who visited the exhibit. The second author transcribed and translated data and artefact materials from Portuguese to English and then both authors engaged in the coding process, identifying initial codes, emergent themes and patterns that allowed us to theorise around our research goals (Creswell, 2013). We conducted this process through thematic coding and constant comparative methods (Patton, 2002).

Building Bridges: Findings and Discussion

In this section, we focus on three major themes that emerged from our analyses: *building connections across formal and informal communities through collaboration, building youth culture,* and *building pathways for change*. These themes emerged from a confluence of voices from "both sides" (museum staff and the school community) and illustrate a broader perspective about the interactions and experiences of different stakeholders.

Building Connections Across Formal and Informal Communities Through Collaboration

Preventing Youth Pregnancy originated from an educational project called *Vale Sonhar* (*It is Worth Dreaming*) developed in state schools of São Paulo (Macedo Guastaferro, 2013) with a focus on teenage pregnancy and STDs. The educational materials developed (predominantly training courses, workshop guidelines and games—all available to teachers), the positive responses of teachers and students about the project, and the visibility the project gained in the media, led to the idea of creating an exhibit about youth pregnancy in the Catavento. One of the exhibit curators and a sex educator of the space, describe how the exhibit came into being and the initial motivation behind its creation:

> And so we thought about an installation based on the [school project] *Vale Sonhar [It is Worth Dreaming]* and the Learning to Live [game]. The *Vale Sonhar* is a project that aims to help in the prevention of pregnancy and it encompasses three workshops. *Learning to Live* is a board game that addresses issues related to AIDS. Actually, it uses the concept of "vulnerability", one of the major axes of the Kaplan Institute and, so, we work with sexual and reproductive rights... teenage sexual and reproductive rights, the concept of vulnerability and the concept of sexuality, in a holistic and broad manner. (Interview, female, 30s, museum curator)
>
> There was an initial project developed with a municipal school in a small community, the *Vale do Ribeira*, where the program had a significant reduction impact ... of pregnancy in teenagers. That was the initial focus of the project *Vale Sonhar [It is Worth Dreaming]*... to decrease the index of youth pregnancy. And then, that material was disseminated as a game, as didactic material, in order to support teachers... with workshops to be developed in the classroom, thinking within national curricular parameters in education, sexuality as a cross-curricular theme... And this space [the exhibit] was created as another space that would attend to that community. (Interview, male, 20s, museum sex educator)

These quotes illustrate the intentionality of the museum staff in their desire to create connections and work with the wider community, including schools. They built on the theme of *sexual orientation* that is part of the Brazilian National Educational Parameters[3] and cuts across science, technology and society. As a cross-curricular topic in Brazilian schools, the focus is primarily on issues such as the human body, gender relations and prevention of sexually transmitted diseases. These issues were deemed as important and highly valued by the visiting teachers as well by the students:

> This is an important topic for their age range. I already know about the project *Vale Sonhar [It Is Worth Dreaming]* but I did not know about the activity in this format. Very cool. (Exit comment card, female, school teacher)
>
> Normal, isn't it? Because you have heard about it [the topic of the exhibit], you get familiar with it in the school. However, there are things you don't know and you feel surprised when you come here. (Interview, visitor, female, 16–20 y, student)

In other words, some of the teachers who visited the exhibit knew its agenda and goals in advance ("I already know about the project Vale Sonhar") and brought with them (positive) expectations about the exhibit. The *Vale Sonhar* project involves a 24 h training course for teachers and principals; support meetings with teachers to help them run workshops that the project promotes; and materials for developing

---

[3] In Brazil, the National Curricular Parameters (PCN) include, for grades 1–4, the cross-curricular theme called *Sexual Orientation*. This theme includes topics such as the human body, gender relations and the prevention of sexually transmitted diseases and AIDS (Secretaria de Educação Fundamental, 1997). In grades 5–8, this theme covers topics such as the body, the continuum of sexuality, gender relations and prevention of sexually transmitted diseases and AIDS (Secretaria de Educação Fundamental, 1998).

the workshops (Instituto Kaplan, n.d.).[4] Additionally, teachers receive the game *Learning to Live* with a focus on AIDS (risks and preventions). With respect to how it all comes together, the curator explained:

> This work [the installation] will complement the one developed in the schools. We are not replacing schools… the teacher who uses the *Vale Sonhar* [*It is Worth Dreaming*] at school … it is not just about him [the teacher] bringing teenagers here and thinking that the work is done. Because the work in the school has a different format, a different approach. (Interview, female, 30s, museum curator)

Previous interactions that the project had established with schools and wider community helped teachers to overcome some of the issues identified by Michie (1998) regarding teachers' lack of time to prepare for a visit, the lack of exchange and facilitation between both worlds (i.e., the school and the museum), and the lack of curriculum connections and relevance. Additionally, these interactions opened up possibilities to continue the work at school beyond the exhibit and beyond the project *Vale Sonhar*.

Although connections with curricular themes were acknowledged as important by both museum staff and teachers, there was an expectation by both of these groups that visitors would be in a space differentiated from school, but that the space and experience therein would complement the work done in the classroom:

> If you ask me "what part [of the exhibit] did you like the most?"—I like the labyrinth, because the labyrinth is able to transpose that concept [different factors affecting youth vulnerability] in a very easy-going, clear way, and it is what makes things different to the methodology used in the school classroom. (Interview, female, 30s, museum curator)
>
> I believe that the goal of our visit was achieved: to offer the students knowledge and an experience completely different from the one that is practiced at school. (Exit comment card, male, school teacher)

The notion of creating learning experiences that are different from those created in the classroom, while at the same time complementing and reinforcing the work done by teachers, is important. Griffin (2004) notes that the combination of resources and strategies between museums and schools provides exciting potential: "for museum and school staff to learn from each other and to learn together" (p. S67).

What is particularly striking in this work is the willingness, by both museums and schools, to engage with complex and controversial socioscientific subject matter, essentially challenging dominant cultural narratives (see Barrett & Sutter, 2006). Science education and traditional science exhibits are often criticised for approaching science as a series of encapsulated topics, void of context, and unproblematic (Bradburne, 1998; Delicado, 2009, Janousek, 2000). In this case, we witness highly charged topics (sex, sexuality and sexual practices) that are alive and well in the Brazilian school context through cross-curricular policy, emerging in the world beyond the classroom replete with social and political nuances. Furthermore, both

---

[4] According to the information provided on the Kaplan Institute's website (http://www.kaplan.org. br/) three workshops are proposed: (1) Identification of the dream, (2) Not all sexual intercourse leads to pregnancy and (3) Getting pregnant is a choice. These workshops include, as the exhibit does, role play and dramatisation.

worlds encourage and support a level of participation (by publics and students) that is iterative, dialogic and open-ended.

Building Connections with Youth Culture

Some of our participants described the exhibit through elements that characterise different aspects of youth culture, such as parties and games:

[In] The second part, the labyrinth, we tried to use a communication approach that could stimulate, that could help, that could trigger adrenaline, that is why a party. And so we have vivid colours... and direct language, as if they were in a party for real, it's what happens at a party. (Interview, female, 30s, museum curator)

I liked a lot the way in which the topic was approached because teenagers need an experience that is close to their reality in order to have significant learning. (Exit comment card, female, school teacher)

The room it nothing more than a real size game. (Interview, male, 20s, museum educator)

The comments above remind us of the notion of *playful pedagogies* described by Buckingham (2003), in other words, pedagogies that engage directly with youth's emotional investment through pleasure and play. Not surprisingly, all of our participants (museum staff, visiting teachers and students), when describing the activities of the exhibit, made reference to role-play and fictional scenarios as an effective way to communicate and connect with young people:

We enter a creative process, the one about dramatisation through the notion of psychodrama... it is important to put the subject [participant] in that dramatisation... The first part [of the exhibit] is the introduction where we ask young people to project into the future, that is the technique... (Interview, female, 30s, museum curator)

I will encourage friends to come because ... it is easy when one person helps the other... I believe that it would prevent many pregnancies... thinking about the future. I found it cool. (Interview, visitor, 18–20 years, student)

The use of drama and role-play can evoke emotion, passion, and speculation about how science and society interact, create memorable experience, and open up possibilities for engaging in moral and ethical discussion (Hughes, 1993; Pedretti, 2002). When asked about the museum's choice of a labyrinth, the curator explained how it represents opportunities for being challenged, for playing, having fun and making decisions: "...the labyrinth is for them to get lost in, to find each other and all that has to be fun, peaceful, and still, you are challenged. This labyrinth has no exit, it is a labyrinth as life is, did you get it?" (Interview, female, 30s, museum curator). While playing, exploring, being separated and reunited, visitors interrogate (individually and collectively) sexual practices, STDs, teenage pregnancy, associated risks and preventions, and the personal, social, economic and ethical consequences of their choices. In short, drama becomes a powerful way of capturing diverse visitor voices and experiences, and enabling discussion among students by creating a safe and trusting space.

Role-play and fictional stories in a museum experience also create opportunities for engaging and exploring youth's emotions by generating "a sense of other" or "being in-the-place-of" (Hodson, 2014, p. 73). Rather than simply "transmitting" information through a one-way communication model, stories and dramatisations can invite multi-directional, open-ended knowledge co-production. The notion of *otherness* also led to important discussions about youth inclusion and groups who are economically disadvantaged and/or marginalised:

> And the idea of a party… we want to bring it … because it was something like, let's try to think that we are in the countryside of São Paulo, what might work there that also could work for them [teenagers] here [São Paulo city] and that will also work in the north east part of the country. (Interview, female, 30s, museum curator)
>
> We are dealing with youth reality, especially the ones from the periphery. (Exit comment card, female, school teacher)
>
> I found it [the exhibit] normal. These are topics that are out there… these are things we already talk about. Nothing made me feel uncomfortable. (Interview, female, 16-20s, student)

Building Pathways for Change

In essence, the exhibition *Preventing Youth Pregnancy* seeks to educate young people, challenge people's points of view, and promote social change and understanding. Museum staff shared the following:

> Our country [Brazil] has lots of preconceptions about diseases, about pregnancy, and so we wanted to go beyond that barrier, to pass over that and to show that these issues are not a joke, this is pretty serious. (Interview, female, 20s, museum educator)
>
> Sometimes there are rigid behaviours and expectations about the roles of women— "women cannot think about sex"—if a young girl has a condom she loses her value as a woman. If she has started to have sex already, there is a discourse about how "loose" she is… So, they [students] bring those issues and we problematise them… "Does it make you vulnerable? How can we confront that?" And we hear what students say … and we try to address through conversation that goal, to break, even a little bit, those ideas. (Interview, male, 20s, museum sex educator)

Our analyses identified *dialogue* and *participation* in a trusting and comfortable environment as central to planting these seeds of change. We note that museum staff had extensive training in order that they could facilitate discussion around sensitive topics with visitors. As one museum curator explained: "There is sexism that appears both in the voice of the boys and the girls and even in the voice of the museum educator if he [sic] is not properly trained" (Interview, female, 30s, museum curator). Both museum staff and visitors commented on the iterative nature of the display and the possibility of building, through facilitation, dialogue and debates, two-way communication and co-learning with visitors:

> So, you have groups and with each group you gain knowledge, a concept… often I say, "Yes, it is the same activity but every group is a group and each group bring its own issues,

their personal issues and you have a lot to learn with them and you have a lot to teach and lots to learn and it is a very intense exchange". (Interview, female, 30s, museum curator).

Being able to create a space for dialogue and play, with an intervention model that would be effective for the group... this is not a completely vertical process but rather one that depends on the participation of the group... This is a space where we are able to welcome them [visitors], receive them, and speak with them, where they can express themselves. (Interview, male, 20s, museum sex educator)

They [the students] liked the excellent way in which the topic was approached: a simulation of a party—having a serious game—and then an open debate. (Exit comment card, female, school teacher)

In addition, staff and teachers' comments suggest that students' active participation, followed by opportunities for reflection, may assist them in reaching informed decisions about their own sexual practices and in developing a sense of agency:

> The experience "experienced" by the students helps them to reflect on the situations that were displayed, much more than the theories in the classroom. All the students, even the shyest ones, actively participated in the experience. (Exit comment card, female, school teacher)

> To have them [the teenagers] using condoms, learning how to use them, critically facing the issue of gender inequalities, calls them to dramatise this experience by acting, doing, being in the place of this experience, reflecting... We always tell them to take condoms with them. Some of them empty the box while others are shy and they take just one. (Interview, male, 20s, museum sex educator)

The dimensions of *dialogue* and *participation* that are embedded in this exhibit contest a *deficit* approach of science communication (see Table 1) typical of many museum exhibitions. The enactment of these dimensions offers visitors a very different kind of experience. Critical exhibitions like *Preventing Youth Pregnancy* reflect more recent approaches to science communication that emphasise content and context and include multi-directional and open-ended interactions, knowledge co-production, discussion about the implications of the information provided (and the choices made around them), debate and praxis (Bucchi, 2008; Levinson, 2010). As Stocklmayer et al. (2010) note: "To draw an analogy with science communication frameworks, we must move from ideas of 'public understanding of science', which have been described as top-down, arrogant and disrespectful, to processes of 'dialogue'" (p. 35). Moreover, critical exhibitions are in tandem with recent calls for a science education that promote and advocate informed decision-making, transactions of ideas, and transformation through civic responsibility (Bencze & Alsop, 2014; Hodson, 2014).

Although the exhibit does not promote political action for change in Levinson's (2010) sense, we suggest that the exhibit creates *conditions* for action (in more personalised ways). *Preventing Youth Pregnancy* promotes learning *about* action (see Hodson, 2014) through: (i) engaging visitors in debates around critical and complex issues (such as youth pregnancy, abortion, prevention of sexually transmitted diseases); (ii) creating opportunities for visitors to experience dissent and conflict (e.g., when preconceptions about sexism are exposed); (iii) connecting visitors with their personal stories and narratives; and (iv) challenging visitors' beliefs and emotions.

We suggest that the exhibit strives to foster *conditions* for action, in the hope that young people will make informed decisions leading to responsible action, and that social norms and practices will slowly shift.

Conclusion: Lessons Learned

Although there are solid arguments in favour of developing controversial exhibitions, the conceptual and practical problems of doing so have been widely reported (Delicado, 2009; Hodder, 2010; Macdonald & Silverstone, 1992; Mazda, 2004; Yaneva, Rabesandratana, & Greiner, 2009). Science centres and science museums typically avoid controversial issues because they are difficult to develop and they tend to impose spatial, sometimes financial, and temporal restrictions—that is, controversies might lose their momentum once the exhibit is opened (indeed, and sadly, the exhibit *Preventing Youth Pregnancy* succumbed to the spatial and financial restrictions often associated with maintaining controversial exhibits, and closed in 2015 after a 6 year run). Moreover, public institutions such as museums and science centres are often stereotyped as places that transmit positive and trustful images of science, reflecting the image of museums as *temples* as described, for example, by Cameron (1971/2004). It is commonly expected (by visitors) that museums avoid ambiguous messages (Macdonald & Silverstone, 1992), which can be an obstacle to presenting different positions and points of view.

The exhibit *Preventing Youth Pregnancy* is no exception. It is a controversial exhibition built on the complex and critical issues of teenage pregnancies, sexuality and sexual practices. It is an intensely personal subject and emotionally loaded for many people. Although teachers, students and museum educators overwhelmingly supported the exhibit, this enthusiasm was not universally widespread. In fact, as the curator explained, one the leading CEOs of the museum questioned, over time, several pieces of the exhibit, including the title (changed from the originally suggested *It Is Worth Dreaming* to *Preventing Youth Pregnancy*); the content; the direct language used; the pertinence of topics such as masturbation; and the age appropriateness of visitors attending (in other words, who should be able to visit). Furthermore, from our interviews and conversations with staff, we learned that museum educators were advised by administration on many instances to avoid the topic of abortion, particularly controversial and sensitive in a predominantly Catholic country.

Although considered controversial by most standards, this exhibit managed to navigate the potentially problematic quagmire inherent in the subject matter, and interact in positive ways with school groups and young people. Our research suggests that this was accomplished by establishing connections and positive interactions early in the process with different community stakeholders, acknowledging and paying careful attention to the role of affect, and integrating diverse models of communication into the exhibition.

Building strong connections between different communities is central to mounting controversial exhibitions that tackle sensitive topics. Collaboration and alignment across communities and community needs can allow for the co-creation of remarkable exhibitions that challenge the status quo (Mazda, 2004) while dealing with important and timely subject matter (in this case, the exhibition topic presented minimal temporal restrictions). This exhibition represented an alliance between the museum (which functioned as host, co-creator and supporter) and the Kaplan Institute (which functioned as the founder and co-creator of the exhibit). They worked side-by-side which allowed the exhibition to exist (advantageously) as an exhibition that was both inside and outside the museum community.

Community connections also help prepare visiting publics and in particular school groups (Chittenden, Farmelo, & Lewenstein, 2004; Pedretti, 2004; Stocklmayer et al. 2010), especially when the connections to cross-curricular topics are strong. In this case study, the exhibit origins are located within an existing school community project that aligns with the Brazilian school curriculum. Teachers were familiar with the community project, and therefore probably more prepared and willing to take their students to this provocative exhibition. This particular exhibition augmented the school experience by allowing for a more participatory, open and negotiated experience. Equally important were the dialogic, participatory and dissent and conflict dimensions of science communication that underlined the narrative around the *creation* of the exhibit. Participating communities adopted models of communication that promoted robust dialogue, debate and ultimately action, rather than a deficit approach (that is, simply transferring knowledge from expert to non-expert), thus maximising the learning potential described by Griffin (2004).

We know that critical exhibitions are often emotionally charged. Pedretti (2004) described the interplay between affect and critical exhibitions as follows: "critical issues-based installations offer something more than simple explication of scientific theories or principles. They strike at the very heart of controversy and debate, and inherently engage visitors by appealing to our intellect *and* our sensibilities" (p. 40). *Preventing Youth Pregnancy* is a topic that involves, among other things, struggles over meaning and morality, power and control. Dissent and conflict are part of the visitor's experience, and as visitors interrogate sexual practices, STDs and associated risks, and prevention associated with teen pregnancy, they are immersed in a labyrinth (both physically and metaphorically) of narratives and choices that have personal, social, economic and ethical implications. The presence of trained sex educators helped encourage and manage sensitive discussions with the students. Such a commitment on the part of museums and science centres to provide skilled facilitators is key to mediating difficult topics and/or situations.

It is clear that the dialogic and participatory dimensions of the science communication model dominate the experiential aspects of the exhibit. However, these dimensions are not at the expense of others. For example, in the *Life* section (incorporated in the tour of this exhibit), there are predominantly more transmissive modes of communication. Similarly the panels and the accompanying text commenting on visitor choices are also primarily transmissive. We argue that these informative pieces are necessary in order to move towards more iterative experiences for visitors. Visitors need to know *something* about, for example, reproductive systems or STDs. Whereas many exhibits begin and end with a deficit model of communication, controversial exhibitions can (and should) use a range of communication dimensions. Museums can (and should) consider the deficit model through the notion of useful knowledge, or, as discussed by Levinson (2010), through the idea of communicating accurate and robust knowledge that may have an important role in empowering deliberation and participation. Therefore, we suggest that controversial exhibitions include, ideally, different dimensions of the communication model, allowing the public to access important information, debate, critically review, reflect, decide, and ultimately take appropriate action on difficult and complex issues.

In conclusion, critical exhibits pose interesting questions about what is presented to the public, the ways in which science centres and museums communicate to and with the public, and how visitors engage with complex and controversial socioscientific material. We are reminded of Cameron's (1971/2004) words, first said over 40 years ago: "Where museums, be they of art, history, or science, have the knowledge and the resources to interpret matters of public importance, no matter how controversial, they are obliged to do so" (pp. 70–71). In spite of the challenges that will inevitably emerge, we advocate that critical and controversial issues should be a part of the science museum *and* school landscape. They carry the possibility for a different kind of learning experience—one that includes participation, dialogue and action—while challenging the status quo.

Appendix

Exemplar panels of the exhibit *Preventing Youth Pregnancy*. The information cited "at the top of the panel" is in response to the choice the visitor made at the previous panel.

Panel 03

(At the top of the panel) If you came from panel 01. Attention! It is a risky behaviour. The condom has to be worn before the penis comes in contact with the vagina.

(At the top of the panel) If you came from panel 17. Congratulations! Always allow some space in the condom for sperm by twisting the end a little bit and letting the air out.

(On the bottom of the panel) You two decided to have anal sex. When the time came penetration was very difficult as the condom had no lubricant. You: (A) Did not take out the condom and suggested to use a water-based lubricant; (B) Took out the condom and continued to have sex.

Panel 08

(At the top of the panel) If you came from panel 01. Congratulations! Any contact between the penis and the vagina is enough for transmitting a sexually transmitted diseases.

(At the top of the panel) If you came from panel 05. Well done! Having sex without using condoms is always risky.

(On the bottom of the panel) Your friend told you that he doesn't use condoms because his girlfriend is on the pill. You decide to do the same: (A) You are not at risk. (B) You are at risk.

Panel 18

(At the top of the panel) If you came from panel 16. Well done! If you lose your condom while having sex; there is risk of having contact with sperm.

(At the top of the panel) If you came from panel 19. Be careful. Spermatozoids move fast and by washing with water you are not going to impede their coming in contact with the egg.

(On the bottom of the panel) Your other half asked you for proof of love: the first time with no condom. You decide: (A) A real proof of love would be to respect your choices and not be at risk. (B) To give that proof of love.

Acknowledgements A special thank you to the staff and visitors of the Catavento museum in São Paulo, Brazil, for being so welcoming and generous with their time. We are also grateful to Mitacs Globalink and SSHRC grant #30124 for funding this research.

References

American Association for the Advancement of Science [AAAS]. (2014). *What is public engagement?* Retrieved from http://www.aaas.org/page/what-public-engagement.

Barrett, M. J., & Sutter, G. C. (2006). A youth forum on sustainability meets the human factor: Challenging cultural narratives in schools and museums. *Canadian Journal of Science, Mathematics and Technology Education, 6*(1), 9–24.

Bell, L. (2008). Engaging the public in technology policy. A new role for science museums. *Science Communication, 29*(3), 386–398.

Bencze, L., & Alsop, S. (2014). Activism! Toward a more radical science and technology education. In L. Bencze & S. Alsop (Eds.), *Activist science and technology education* (pp. 1–19). Dordrecht, The Netherlands: Springer.

Bradburne, J. M. (1998). Dinosaurs and white elephants: The science center in the 21st century. *Museum Management and Curatorship, 17*(2), 119–137.

Bucchi, M. (2008). Of deficits, deviations and dialogues: Theories of public communication of science. In M. Bucchi & B. Trench (Eds.), *Handbook of public communication of science and technology* (pp. 57–76). New York, NY: Routledge.

Buckingham, D. (2003). Media education and the end of the critical consumer. *Harvard Educational Review, 73*(3), 309–327.

Cameron, D. (1971/2004). The museum, a temple or the forum. In G. Anderson (Ed.), *Reinventing the museum* (pp. 61–73). New York, NY: Altamira.

Chittenden, D., Farmelo, G., & Lewenstein, B. V. (2004). *Creating connections: Museums and the public understanding of current research*. Walnut Creek, CA: AltaMira.

Creswell, J. W. (2013). *Qualitative inquiry and research design: Choosing among five approaches* (3rd ed.). London, UK: Sage.

Delicado, A. (2009). Scientific controversies in museums: Notes from a semi-peripheral country. *Public Understanding of Science, 18*(6), 759–767.

Diamond, J., Luke, J. J., & Uttal, D. H. (2009). *Practical evaluation guide: Tools for museums & other informal educational settings* (2nd ed.). Lanham, MD: AltaMira.

Durant, J. (2004). The challenge and opportunity of presenting 'unfinished science'. In D. Chittenden, G. Farmelo, & B. V. Lewenstein (Eds.), *Creating connections: museums and the public understanding of current research* (pp. 47–60). Walnut Creek, CA: AltaMira.

Einsiedel, A. A., & Einsiedel, E. F. (2004). Museums as agora: Diversifying approaches to engaging publics in research. In D. Chittenden, G. Farmelo, & B. V. Lewenstein (Eds.), *Creating connections: Museums and the public understanding of current research* (pp. 73–86). Walnut Creek, CA: AltaMira.

Gascoigne, T., Cheng, D., Claessens, M., Metcalfe, J., & Schiele, B. (2010). Is science communication its own field? *Journal of Science Communication, 9*(3), C04.

Griffin, J. (2004). Research on students and museums: Looking more closely at the students in school groups. *Research on Students and Museums, 88*, S59–S70.

Hodder, A. P. W. (2010). Out of the laboratory and into the knowledge economy: A context for the evolution of New Zealand science centres. *Public Understanding of Science, 19*(3), 335–354.

Hodson, D. (2014). Becoming part of the solution: Learning about activism, learning through activism, learning from activism. In L. Bencze & S. Alsop (Eds.), *Activist science and technology education* (pp. 67–98). Dordrecht, The Netherlands: Springer.

House of Lords. (2000). *Science and technology*. Third report. Retrieved from http://www.publications.parliament.uk/pa/ld199900/ldselect/ldsctech/38/3801.htm

Hughes, C. (1993). *Perspectives on museum theatre*. Washington, DC: American Association of Museums.

Instituto Kaplan. (n.d.). *Historia* (History). Retrieved from http://www.kaplan.org.br

Janousek, I. (2000). The 'context museum': Integrating science and culture. *Museum International, 52*(4), 21–24.

Levinson, R. (2010). Science education and democratic participation: An uneasy congruence? *Studies in Science Education, 46*(1), 69–119.

Lewenstein, B. (2003). Editorial. *Public Understanding of Science, 12*, 357–358.

Lincoln, Y. S., & Guba, E. G. (1985). *Naturalistic inquiry*. Beverly Hills, CA: Sage.

Macdonald, S. (1998). *The politics of display: Museums, science, culture*. New York, NY: Routledge.

Macdonald, S., & Silverstone, R. (1992). Science on display: The representation of scientific controversy in museum exhibitions. *Public Understanding of Science, 1*, 69–87.

Macedo Guastaferro, C. (2013). *Adolescência, Gravidez e Doenças Sexualmente Transmissíveis (DST): Como os adolescentes enfrentam estas vulnerabilidades?* [Teenage, pregnancy and sexually transmitted diseases (STD): How do teenagers face those vulnerabilities?]. Unpublished Masters dissertation, Universidade Federal de São Paulo.

Mazda, X. (2004). Dangerous ground? Public engagement with scientific controversy. In D. Chittenden, G. Farmelo, & B. V. Lewenstein (Eds.), *Creating connections: Museums and the public understanding of current research* (pp. 127–144). Walnut Creek, CA: AltaMira.

Michie, M. (1998). *Factors influencing secondary science teachers taking field trips*. Darwin, Australia: Northern Territory Department of Education.

Patton, M. Q. (2002). *Qualitative research and evaluation methods* (3rd ed.). Thousand Oaks, CA: Sage.

Pedretti, E. (2002). T. Kuhn meets T. rex: Critical conversations and new directions in science centres and science museums. *Studies in Science Education, 37*(1), 1–41.

Pedretti, E. (2004). Perspectives on learning through research on critical issues-based science center exhibitions. *Science Education, 88*(1), S34–S47.

Pedretti, E. (2012). The medium is the message. In E. Davidsson & A. Jakobsson (Eds.), *Understanding interactions at science centers and museums: Approaching sociocultural perspectives* (pp. 45–61). Rotterdam, The Netherlands: Sense.

Pedretti, E., & Dubek, M. (2015). Critical issues-based exhibitions. In R. Gunstone (Ed.), *Encyclopedia of science education* (pp. 236–238). Dordrecht, The Netherlands: Springer.

Pouliot, C. (2009). Using the deficit model, public debate model and co-production of knowledge models to interpret points of view of students concerning citizens' participation in socioscientific issues. *International Journal of Environmental and Science Education, 4*(1), 49–73.

Schiele, B. (2008). On and about the deficit model in an age of free flow. In D. Cheng, M. Claessens, T. Gascoigne, J. Metcalfe, B. Schiele, & S. Shi (Eds.), *Communicating science in social contexts: New models, new practices* (pp. 93–117). Dordrecht, The Netherlands: Springer.

Secretaria de Educação Fundamental. (1997). *Parâmetros curriculares nacionais: apresentação dos temas transversais, ética* [National Curricular Parametres: Introduction to cross-curricular themes, ethics]. Brasília: MEC/SEF. Retrieved from http://portal.mec.gov.br/seb/arquivos/pdf/livro081.pdf

Secretaria de Educação Fundamental. (1998). *Parâmetros curriculares nacionais: terceiro e quarto ciclos: apresentação dos temas transversais* [National Curricular Parametres: third and fourth cycle: introduction to cross-curricular themes]. Brasília: MEC/SEF. Retrieved from http://portal.mec.gov.br/seb/arquivos/pdf/ttransversais.pdf

Soren, B. J., & Armstrong, J. (2014). Qualitative and quantitative audience measures. In G. D. Lord & B. Lord (Eds.), *The manual of museum exhibitions* (pp. 58–66). London, UK: The Stationary Office.

Stocklmayer, S., Rennie, L. J., & Gilbert, J. K. (2010). The roles of the formal and informal sectors in the provision of effective science education. *Studies in Science Education, 46*(1), 1–44.

Trench, B., & Bucchi, M. (2010). Science communication, an emergent discipline. *Journal of Science Communication, 9*(3), CO3.

Yaneva, A., Rabesandratana, T. M., & Greiner, B. (2009). Staging scientific controversies: A gallery test on science museums' interactivity. *Public Understanding of Science, 18*(1), 79–90.

Encounters with a Narwhal: Revitalising Science Education's Capacity to Affect and Be Affected

Steve Alsop and Justin Dillon

Abstract Informal science education has always placed considerable importance on the emotional and physical aspects of learning science. In contrast, however, science education in formal contexts and in research tends to favour largely disembodied accounts of both teaching and learning. These commonly place an emphasis on knowledge, language and culture more than experiences, embodiments and affect. In this chapter, we explore teaching and learning of science as an embodied phenomenon. This hinges on a body's capacity to affect and be affected. Learning science, in these terms, is learning to be affected by science as well as learning to affect science. We take efficacious pedagogy as a purposeful framing of different encounters *enhancing* this capacity. We apply this unusual perspective to describe the first author's pedagogical entanglements with a preserved narwhal (within a particular museum setting). We conclude with considerations of how these encounters—and more generally science education (theory and practices)—might learn to 'live better' with charismatic endangered creatures in the era of the anthropocene marked by rapid ecological declines. Our general argument is that we need much more talk of embodied affects in science education and this can have far reaching consequences for science education in all settings.

Keywords Museum · Affect theory · Spinoza · Biodiversity · Narwhals · Pedagogy · Embodiment · Gilles Deleuze

S. Alsop (✉)
York University, Toronto, ON, Canada
e-mail: SAlsop@edu.yorku.ca

J. Dillon
Exeter University, Exeter, UK
e-mail: J.S.Dillon@exeter.ac.uk

© Springer International Publishing AG, part of Springer Nature 2018   51
D. Corrigan et al. (eds.), *Navigating the Changing Landscape of Formal and Informal Science Learning Opportunities*,
https://doi.org/10.1007/978-3-319-89761-5_4

Encountering a Narwhal in The Royal Ontario Museum, Toronto, Canada
Author's picture

A Museum Visit

The Royal Ontario Museum—or the ROM, as it is referred to locally—has opened a
new exhibit within its impressive Schad Gallery of Biodiversity. The focus is a rather
sombre contemporary theme—*Life in Crisis*. A central specimen in the exhibit is a
preserved narwhal. This exhibit provides the context for our analysis. The chapter starts
with the first author describing a visit to the museum, framed by a particular perspective
in affect theory. The visit is represented in three encounters: (i) angles of arrival; (ii)
fragility and loss; and (iii) awe and awkwardness. A photograph accompanies each
encounter. These were taken during the visit and seek to further illustrate the accompanying first person narratives. We then offer a series of reflections on the visit, exploring
'teaching', and 'learning' 'and 'affect', and the theme of the edited collection, changing
landscapes of formal and informal science learning opportunities. The chapter offers a
less common way of conceiving of science education, embracing the subjective,
embodied, emotional and relational dimensions of our experiences and awareness.

(i) Angles of arrival

As I walk along the long corridor into the gallery, I become consciously aware of
my 'angle of arrival' as well as the 'mood of the museum'. As the affect theorist

Kathleen Stewart (2010) writes, everything "depends on the feel of the atmosphere and the angle of arrival" (p. 337). My angle of arrival is marked by my experiences, knowledge and intentions. We bring agendas with us in our educational journeys, shaping our awakenings and encounters in some ways rather than others. My recent research has been exploring affect, largely informed through a particular definition of a bodily capacity *to affect and be affected*. This approach follows a theoretical trajectory associated with Gilles Deleuze's (1988) reading of Baruch Spinoza and William James, and more contemporary 'affect theorists' including Melissa Gregg and Gregory Seigworth (2010), Sarah Ahmed (2010) and Brian Massumi (2015). In this visit, I carry this definition and these scholars with me, contemplating my teaching and learning as a situated, embodied feeling process.

Affect theorists often write about arrivals as orientations, of being moved toward or away by something near, and thereby allowing other things to fade (Gregg & Seigworth, 2010). Other theorists prefer different terms, including fluctuating 'intensities' (Betelsen & Murphie, 2010), 'bloom spaces' (Stewart, 2010) or 'shimmers' (Barthes, 2005). Encounters in this sense are embodied, feeling-orientated and relational. They are continuous beginnings: how we dispose, awaken or orientate (different affect theorist use different words) ourselves to encounter affects how and what we encounter. How we encounter, in turn, affects our capacities to dispose, awaken and orientate. In this manner, the promises of educational experiences are in their evolving capacities to direct us toward something different and generative, whether it is knowledge, practices, objects, sensations, feelings or new ways of being in science education. Through such processes some things, some movements and some utterances now stand out; they gain intensities of feeling—they shimmer or move us in some way. Our capacities, thereby, can become more discerning. We gain abilities to tune in, noticing what was previously eclipsed, or perhaps only partly in shade.

In the distance, the narwhal is now in sight.

Getting closer to the narwhal involves navigating a busy corridor full of glass cabinets, specimens, signs and hordes of school children frantically being chaperoned by hassled teachers. I immediately become aware that this is going to be a rather loud encounter with multiple human and preserved non-human bodies mingling together—interacting in very limited space. Although I try to shut out the hustle and bustle, it is always present, and at times it is exhausting. The room seems oxygen-less. Seemingly unaware, a group of children pass-by enthusiastically chanting in unison: "*Narwhals, Narwhals swimming in the ocean. Causing a commotion 'Cause they are so awesome*". I later find out that this is a rather addictive children's television song.[1] I should admit that I had no idea at the time.

It is impossible to fully capture a sense of atmosphere and ambience in words. These phenomena seem to escape such modes of representation. Perhaps they only become discernable when they are felt in visitors' energies and fatigue. Navigating this museum seems to require a particular frame of 'body/mind'—a willingness to share and an obstinacy to stay in one place for a period of time. I am so conscious

---

[1] The Narwhal Song is available on YouTube, see: https://www.youtube.com/watch?v=anM1N5oN-OM. (Last accessed, Dec 2015). Warning: It is incredibly infectious.

of movement. Like the exhibits, however, I remain stationary. In contrast, most other visitors seem in constant motion, darting and dancing from one experience to another. The Brazilian educator Paulo Freire once described pedagogy as a dance of knowledge. I get a sense of what he means. I look up and take a photograph (carefully contrived to exclude others, in compliance with standard human participant research protocols).

Entering the *Life in Crisis* exhibit in the Schad Gallery of Biodiversity, Royal Ontario Museum

There is my educational target. Overhead. There is something 'bizarre' (perhaps 'unreal' or 'unsettling' might be better words here) about encountering a six-feet long aquatic mammal, with a three-foot tusk, suspended from a ceiling in a downtown Toronto museum. It lurks so silently and so lifelessly above, forcing uncomfortable viewing as you turn upwards and stare into its smooth and shiny underbelly. I start to attend to its shape and form. It is impressively sleek and streamlined, conveying a sense of speed and agility. It looks like a dart. Although strikingly large and entirely static, it is strangely evasive, positioned in such a way to make it difficult to get a satisfactory and comfortable view. I openly wonder if this is part of the intended curriculum, and long for a better perch.

Rachel Poliquin writes evocatively of natural history museums as "breathless zoos". Taxidermy, she says, is like storytelling, deeply marked by human longing. Her historical analysis traces taxonomic practices, marking a shift in early eighteenth Century France in which taxidermy reoriented from representing "wonderment" to representing "beauty", focusing on conveying "visual pleasures by seeking to conform to aesthetic principles of the day" (p. 57). I wonder what the aesthetic curriculum of this narwhal is? Is it longing for wonderment, or for beauty, or for something else entirely? In our interactions, I wonder, what we are conforming to (and not)? What type of dance of knowledge are we performing?

Specimens continue to play a central role in science, as numerous science and technology scholars attest. In this respect this specimen is likely to be more scientifically real than most objects encountered in any school science classroom. Yet despite this authenticity, its pedagogy is quite difficult to pin-down. I am left with several questions: What does the presence of a narwhal specimen add to this exhibit (which a 2D symbolic representation cannot adequately convey)? What is gained by our bodily encounters? In what sense does this narwhal teach? What kind of teaching object is it? What kind of subject is it teaching? Even more fundamentally, in what sense is this even science education at all?

Donna Haraway (2008) and many other authors in science studies teach us that "objects are not simply 'raw material' to the potency and action of intentional others" (p. 262). Objects, in this sense, have agencies and associated "thingness" or "vibrancy" (Bennett, 2010). There is modest research in science education on how museum specimens (or any other material objects in science education) teach. Most of our focus is on how students' learn. Teaching, if mentioned at all, is mostly conceived of as a human-to-human affair, in the tradition of schools and other formal institutions of education.

I stand on my tiptoes to get a closer look.

(ii) Fragility and loss

How do objects acquire meanings? How do meanings acquire objects? These issues can be lost when we teach that scientific meanings simply represent or mirror natural phenomena. This is an ontology underwritten by linguistic representationism. In the gallery, I swiftly become aware that the specimens are arranged by a concept: climatic region. These glass cases are stuffed with preserved Arctic fauna and flora, natural treasures from the North. This collection comprises birds, an Arctic fox, a polar bear with a seal, plus our narwhal and some other animals. It is intended to represent a sense of an entire Arctic ecosystem. This is the story gathering this motley crew together in their "afterlives" (Alberti, 2010). It is the "explicit curriculum", to use Elliot Eisener's famous term.

This curriculum, however, involves a considerable stretch of the sensory imagination. You need to allow yourself to really play along with the story line. After all, this is an exhibit of inanimate stuffed individuals. As I stare at these specimens, I get a sense of their physical, materially embodied presence. They are frozen, permanently set in rigid bodily poses. If these creatures were to actually encounter one another in real life, things would of course be much more expressive and altogether more animated. Ecosystems are, perhaps above all else, dynamic. In this respect, these breathless bodies seem largely out of kilter. Matter and meanings, objects and ideas now dissociate—the material-curriculum (the specimens and their assemblage) and the accompanying conceptual-curriculum (the notion of an ecosystem) seem somehow to be shifting out-of-joint.

I notice the large red banner on the back of the display cabinet. It is impossible to miss. The sign makes it authoritatively clear that this exhibit is not only a story of biodiversity, but also a warning of disastrous species loss and extinction.

HUMANS are causing both the extinction of individual species and the destruction of whole ecosystems

Temperatures are rising almost twice as fast in the Arctic as on the rest of the planet

Due to global warming, scientists predict the Arctic could be ice-free in summer by 2030

If current warming trends continue unabated, scientists believe that two thirds of the Polar Bear population will be extinct by 2050

The Arctic is an unnatural "sink" for toxic chemicals from the industrialized south

The large red sign: The main text in the exhibit

I read the sign and then turn to the narwhal. In doing so, I sense little more than ambiguity—an affect of loss but at the same time no loss. I am not questioning the fragility and vulnerability of Arctic ecosystems. I do deeply care. But, what should I be feeling here and right now? What should my bodily sensations or reactions be? How should I be affecting/affected? After all, this narwhal doesn't exactly appear in crisis. On the contrary, it seems aesthetically beautiful, sleek, agile and athletic. It is an ideal specimen: geometrically balanced, almost perfectly symmetrical and entirely unblemished. It effortlessly conveys beauty and agility, rather than mourning, vulnerability or loss. Perhaps the point is that this specimen was created with other stories and longings in mind, at another point in time. It has now found itself caught up with a red sign and a contemporary genre of ecological crisis. This positioning raises important questions about feelings and affectations and what it might mean for science education to re-orientate its practices to contemporary ecological conditions.

Robert Kirkman (2007) is not alone when he asks us to care more about climate change and loss of biodiversity. His analysis builds a convincing case for the importance of experiencing human vulnerability to jolt us into change. An "objectivist account of vulnerability", in terms of a theoretical threat, he argues, is simply not enough. What we need are accounts that "we feel in our bones" (p. 20). We need a deep-felt sense of our own vulnerability to make sense of the vulnerabilities of others. Kirkman's choice of theorist is Maurice Merleau-Ponty and his unfinished book, *The visible and the invisible.* In this work, Merleau-Ponty lays out plans to conceive of science as embodied action, an argument built on a notion of "flesh". At one point, Kirkman (2007) cites Merleau-Ponty: "I perceive the world only because of the flesh of my body intertwines with the flesh of the world" (p. 21). This talk of flesh seems far more symbolic of a movie trope than a science lesson—perhaps a movie along the lines of *The Return of the Living Dead* or *Night at the Museum.*

Our focus is on a pedagogical meeting between a specimen and science educator. Donna Haraway (2008) writes about multispecies "contact zones" and "syntactically and materially, worldly embodiment" (p. 250). Although I am unable to feel anything intense, or anything "in my bones" (in Kirkman's sense), I am certainly aware of the material presence of this object. The specimen is quite captivating; it demonstrably grabs my attention for some protracted period of time. It also orientates this attention, drawing me into some features more than others.

Sarah Ahmed (2010), a high profile affect theory scholar, writes of some objects being "sticky". Affect in these terms, she argues, "is what sticks, or sustains or preserves connections between ideas, values, and objects" (p. 30). This object certainly captures my gaze; it is pedagogically sticky thereby affecting me in particular ways. It is captivating/I am captivated. The specimen, in these terms, has affective value; it has gained some, or been granted some, capacity to affect. What is pedagogically puzzling, perhaps, is a question of emphasis. I like to think that I am doing so much more than this narwhal, but what pedagogical work might be associated with this specimen? What is it capable of, and not? What role does this dead inanimate specimen have in shaping/magnifying/mediating/affecting (there are lots of possible words here) my feelings of captivation?

I become aware that I am standing in a room surrounded by a variety of stuffed dead things and being provoked to conceive of flesh, agency and species loss. Awkward. At least there are still lots of excited and energetic children running around, and I reassure myself that it's not exactly easy to get bodily in tune with extinction and these larger existential questions. I am reminded of Samuel Alberti (2010) who cites Henry David Thoreau:

> I hate museums; there is nothing so weighs upon my spirits. They are catacombs of nature. One green bud of spring, one willow catkin, one faint trill from a migrating sparrow would set the world on its legs again. The life that is in a single green weed is of more worth than all this death. They are dead nature collected by dead men. I know not whether I must muse most at the bodies stuffed with cotton and sawdust or those stuffed with bowels and fleshy fibre outside the cases. (p. 5)

I know what he means. Perhaps, with the best will in the world, there is only so much that a stuffed specimen can hope to represent and/or convey. The philosopher Hans Jonas (1982) claims that 'life' can only truly know 'life'. Given the profound significance of specimens in the development of scientific knowledge, I openly wonder what I might have been missing, and what I might hope to recover in navigating formal and informal educational contexts. For instance, what might it mean to be alive (or not) in science and in science education (Alsop, 2011)? I still don't think this is as straightforward a question as it can at first seem.

(iii) **Awe, intimacy and awkwardness**

Narwhals, of course, arrive with familiar pre-packaged stories that, over time, they have been involved in shaping. Narwhals are magical, charismatic, exotic mega-creatures. Found in the north of Canada, they are referred to as *Qilalugaq Tugaalik* in the local Inuit language of Inuktitut. These creatures are deeply immersed in myth, mystery, wonderment and awe. They have been the focus of much recent scientific and cultural history research. Although not directly encountered by most Canadians, narwhals are often romanticised; representations of the wild, exotic, true North. For some they are entwined in Canadian cultural identity, for others not. Jens Rosing (1999) in a publication entitled *The unicorn and the Arctic Sea* traces narwhal cultural history over a staggering period of 4000 years. Throughout history, narwhals have become entangled with myths of unicorns, seen as possessing truly magical powers. Rosing, at one point, evocatively describes a real-life game-of-thrones: King Frederik III of Denmark (1648–1670) ordered a throne to be built entirely out of narwhal tusks. Although he died before the project was finalised, his son, Christian V got to sit on the impressive 342-pound tusk chair, thought to carry magical powers.

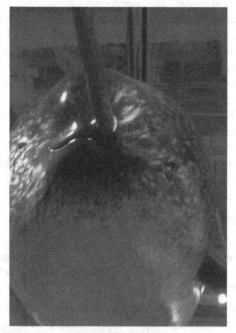

Mystery and wonderment: Encountering a unicorn or a whale?

Knowing this history affects your gaze; it makes you more open in some ways and turns your attention to some particularities and specifics. Knowledge, of course, has the capacity to frame our encounters. We learn to be affected in different ways—becoming more discerning and articulate, attending to some things rather than others. I focus intently on the extraordinary tusk. It is actually a giant left canine tooth, a fact that becomes apparent when you examine it, awkwardly protruding from the left side of the specimen's month (as the picture shows). Given the tusk, I infer that this narwhal is male. Females only occasionally grow tusks. A narwhal's body length typically ranges from 13–18 ft. So this is a small narwhal, probably a young male. The twist of the tusk/tooth is usually clockwise, which it is in this case. It contains thousands of nerve cells and there is still much debate over its function and utility (WWF, 2015). As I focus on the base of the tooth, I am struck by affective qualities of 'awe' but also 'awkwardness'. The awesome tusk being out of symmetry seems inelegant and discomforting in some ways.

I ask myself, are there differences between encountering a narwhal as a unicorn or a whale? I think there are. The affections of mystery and wonderment seem to fade (or at least become secondary) when the narwhal becomes *Monodon monoceros*, a medium-sized, pale-coloured whale with a large 'tusk'. The physical demise of this narwhal is simultaneously the birth of a universalised scientific object. Samuel Alberti writes provocatively of "chunks of landscape ripped asunder and transported to urban locales" (p. 4). This is a story of knowledge in transit, of natural history in the making. But, as I look up and stare, there is still something wondrous and awesome about this creature, even in this most rudimentary form. It is more than a neutral chunk of the world; there is still a residual unicorn lurking somewhere.

I am once more reminded of the role that taxidermy and natural history museums have in the development of science (see Star and Griesemer (1989), for instance). The point is that some animals just move too quickly, or too slowly, to be examined closely. Others escape our grasp because they are too large, or too small, or simply too obstinate to pin-down in one-way or another. As Bruno Latour (1983) persuasively argued, scientists need objects that are "mobile" and also "immutable"—objects that are presentable, readable and can be combined with one another. Much like the primary school practices of 'show-and-tell' or secondary school laboratory experiments and demonstrations, science involves acts of convincing others through manageable material entities that are predictable, engaging and plausible.

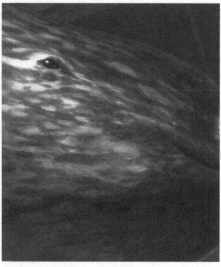

Coming eye-to-eye with the narwhal

Pushing things a little further, I look up and come eye-to-eye with this narwhal. What am I sensing? How am I being affected? What am I becoming affected by? What does it even mean to come eye-to-eye with a dead whale, a deceased stuffed thing? It borders on the mystical and the macabre. I gaze into what becomes recognisable as a glass substitute eye. Rachel Poliquin's (2008) research documents visitors describing taxidermy as "eerie" and "haunting" (p. 123). There is a familiar sensation of "the eyes following me". I get a real sense of this—a feeling of the Freudian uncanny, or of a ghostly otherworld. Such experiences serve, perhaps above all else, to "evince taxidermy's provocative visceral presence" (p. 130). I affect and this specimen affects me, there is a dynamic relationship emerging here, however unlikely it might at first seem.

In so many ways, however, taxidermy specimens are poor imitations of the real thing. They are human re-constructions of the natural; dynamic blends of carcases and culture. They are, perhaps, nothing more than impoverished impostures; cartoon replicas of reality, underwritten by sensitivities and preoccupations of the past (Poliquin, 2011). Encountering a narwhal in the Arctic, or any other whale in the wild, is likely to be so different. Indeed many of us have been fortunate enough to experience real-world whale watching, in which, turning to the words of Donna Haraway (2008), "we sense that inside this other body, there is someone home, someone like ourselves" (p. 236). In whale-watching trips many describe deeply affective entanglements with sentient non-human cetacean others (see Staus & Falk, 2013). By any measure, coming eye-to-eye with this museum specimen is less intense, way more predictable and far less emotionally moving. Perhaps this is the point; it is a 'scientific' specimen designed to be neutrally objective—a single specimen that represents all narwhals. In so doing, it loses any sense of its individual life. Now I leave with questions of affective authenticity: In what ways might this educational encounter even be compared with a real encounter with a narwhal? Is there anything real here at all? What is it missing? Am I being deceived at this moment of objectification and loss of intimacy?

My plan was to spend about an hour or so in the exhibit, *thinking* though affect, taking field-notes and photographs in preparation for this book chapter. However, after about 30 min or so, I start to feel somewhat socially awkward and out-of-place. I become conscious that I am the only person spending a protracted period of time at this or any exhibit. Transience seems the dominant theme. Philip Payne and Brian Wattchow (2009) write of the importance of "slow pedagogy". I start to wonder when such pedagogy becomes loitering? Perhaps this is the point. Educational settings are rarely structured for dwelling, but rather for rapid-fire, move-on encounters. I wonder how this exhibit might be redesigned with slower, more meandering goals in mind. As another school group barges past, I head for the café in search of coffee and cake.

Debriefing the Narwhal

What might these museum reflections offer science education? What might we learn from such individualised, idiosyncratic encounters with a preserved narwhal? How might these descriptions add to the theme of this edited collection—navigating formal and informal learning opportunities?

In answering such questions there is always a danger of over-generalising. The descriptions of this museum visit do not purport to represent a comprehensive treatment of museum-based science education. How could they? They are completely immersed in the intentions, experiences and expertise of a single visitor and exhibit. They are underwritten by particular desires and specific knowledge of narwhals, science education, affect theory, and science studies. They are deeply personalised. But this, we suggest, is part of both their purchase and appeal. These narratives make no claim to objectivity, in the sense of definitely representing the educational efficacy of this exhibit. We recognise that this is an important consideration, but we offer these encounters here for slightly different purposes. We offer them as a pause for thought about *subjectivities* and the roles that feelings play within all science and education.

To better understand what makes this exhibit work, we posit, needs greater attention to unique 'angles of arrival' and 'embodied capacities of fully immersed visitors', as well as 'atmospheres' and 'moods' of museums, and the work of those 'sticky' inanimate objects that hang from ceilings in their after-lives, wrapped-up in stories of exotic Northern ecosystems and biodiversity loss. We need to attend more closely to feelings of awe and awkwardness, and fragility and loss. Our analysis is thus an invitation to better understand science and education through attending more closely to emerging affective relationships and their underpinning processes of formation. To think with affect, in this way, is to embrace education as a process—a process that, to use William James's phrase, is always "in the making". This approach contrasts sharply with a way of thinking about education as acquiring or constructing 'what is out there' in a 'static', detached, rational and objectified sense.

Our story of the narwhal is thus a story of educational encounters and their capacities to intensify and mediate; their abilities—returning once more to our opening conceptual guide—to 'affect and be affected'. This is an understanding of science education that is always on-going, in-flux, and enfolding both meanings and contexts. Indeed, the processual nature of these discussions distinguishes them from analytic ways in which traditional disciplinary knowledge is often built. We are now exploring how learners become part of exhibits (and not) and how exhibits simultaneously become part of learners (or not). This is not simply about a narrow conceptual outcome.

This is very much a story of feelings and emotions, but not in a traditional sense of a Cartesian mind/body, cognition/emotion duality. In this type of analysis, feelings and emotions are traditionally represented as separate, mediators of cognition. In contrast, our analysis has sought to embrace the inseparable, in-dissociable nature of cognition, feelings and emotions. Following Baruch Spinoza's monism, affect theorists argue that all our worldly encounters are co-constitutively shaped by our embodied (including mind) arousals and awakenings. These include, of course, our hopes, our fears, our joys and frailties—those anticipations, ambiguities, aspirations and hesitations that shape why, how and what we seek to achieve, or not. These affectations, and their associated intensities of feeling, drive our explorations within the world, whether in science, or education, or any other human endeavour. Learning science, in such terms, is not solely about acquiring a conceptual outcome, but is a subjective condition of more fully embracing and realising our ongoing and emerging sensory, embodied participation.

We can now imagine walking into a classroom, or visiting a museum, with our attentions redirected, our angles of intensities shifted, thereby being oriented towards new encounters and experiences. A central theme in such an analysis is how one opens or relates to 'the other', and thereby how this 'other' opens or relates to one. Embracing such intermediaries is provocative and revealing. It is also controversial. The point is that the ways that we dispose ourselves to encounter affects how and what we encounter. Indeed, it is possible to consider a plethora of ways or states in which one might encounter this narwhal and thereby come to affect and be affected by it in differing ways. John Law (2004) writes evocatively about different ways of knowing within research methods:

> Perhaps we will need to know them through the hungers, tastes, discomforts, or pains of our bodies. These would be forms of knowing as embodiment. Perhaps we will need to know them through 'private' emotions that open us to the worlds of sensibilities, as emotionality or apprehension. Perhaps we will need to rethink our ideas about clarity and rigour, and find ways of knowing the indistinct and the slippery without trying to grasp and hold them tight. Here knowing would become possible through techniques of deliberate imprecision. Perhaps we will need to rethink how far whatever it is that we know travels and whether it still makes sense in other locations, and if so how. This would be knowing as situated inquiry. Almost certainly we will need to think hard about our relations with whatever it is we know, and ask how far the process of knowing it also brings it into being. (p. 2)

In such ontological terms, one might conceive of a multiplicity of narwhals each brought into existence through differing orientated encounters. Of course, the narwhal specimen is not absent within these discussions; its material presence might support or diminish—reflect, refract or diffract (see Barad, 2003)—some of these orientations, or not. Some will seem appropriate, while others wildly fanciful and out-of-place. However, the key point is that we have escaped notions of teaching and learning as 'objectively representing', of telling and accounting for a narwhal as a solitary language story.

In this manner we join with contemporary theorists in science-technology-society, including Bruno Latour, Karen Barad, and Jane Bennett (and many others) who suggest that material matter has agency. *Materiality* is the basis of an increasing number of contemporary theories including: Actor Network Theory (ANT)—see Bruno Latour (2005); Agential Realism (Karen Barad, 2003);

Post-Phenomenology (Don Ihde, 1990); and Jane Bennett's (2010) *Political Ecology of Things*. As a result we leave wondering about *material pedagogy*, the multiplicity of ways in which different scientific objects (narwhals, Bunsen burners, gas taps, DNA models, test-tubes, van de Graaff generators, whiteboards, tables, desks and chairs, and so many other things of course) affect and are affected by teachers and learners. Tobias Roehl (2012) offers these related questions about science classrooms: "How are students performed by material objects? What kind of student subject emerges in the engagement with the material object? How are students bodily affected? What kind of school lesson emerges? How is classroom discourse transformed via the object?" (p. 54). The general point is that the material agencies of pedagogical objects call for greater attention in both our research and pedagogical practices.

Navigating Changing Landscapes

Let us now turn directly to this edited collection's central theme. There has been a lasting tradition in science education of drawing distinctions between formal and informal learning. A significant feature of this work is demarcating some of the differences that these *contexts* present. This type of analysis is predominately structural—bringing attention to how different cultural settings (e.g., schools/universities/colleges versus museums/science centres/media/others) seek to account for science education in quite different ways. As others in this edited collection point out, such distinctions seem somewhat dated now. Although they were once helpful in bringing groups together with common interests for change (especially those with a shared interest in informal settings), they can also be restrictive and limiting. By focusing on structure, for instance, they seem to mask complexities of learners' agencies, foreclosing the possibilities that learners might approach informal contexts with formal intentions and vice versa (see chapter "Using a Digital Platform to Mediate Intentional and Incidental Science Learning" by Cathy Buntting, Alister Jones and Bronwen Cowie). Through an emphasis on language (symbolic representations) and different learning outcomes they can also overlook key features of learners' embodied 'presence' in learning environments.

This chapter has focused attention on a learner and the ways in which this learner's capacities to 'affect and be affected' might be brought more fully into play in navigating changing landscapes in formal and informal science education. In offering a relational view of learning, this analysis seeks to traverse structures and agencies, linking and dynamically relating educational objects and subjects. To think with affect is to navigate learning opportunities in different ways, by allowing learners capacities to *affect* and *be affected in* science education settings and contexts.

Moreover, we recognise subjective dimensions of our knowing and awareness—those subjective, embodied, situated, feelings that underwrite all teaching and learning. Such dimensions are so easily discouraged by powerful voices of tradition forcing our attentions toward standardised knowledge and representative practices.

There is much to learn from a museum, in which intensities of feelings associated with mood, objects, movements and orientation seem so central.

So, where to now? What do we need as science educators and researchers in pursuit of transformations and deep-rooted change? What might it mean to scale-up these discussions? Perhaps the best starting point for change in science education is with our consciousness, our perceptions, our feelings and our bodies, by bringing new and unexpected patterns into our science openings and experiences. In order to lead others, we need to nurture our capacities to sense, to think, to feel, to discern science and education in different ways and for different purposes. It is at this point that we can invite others into the promises of our experiences by nurturing their capacities to share our encounters in science education.

This, perhaps, is the bigger picture—it is about amplifying awareness and shifting perceptual dials, with prospects of realising different and increasingly more discernable and describable encounters and experiences. It is the first author's hope that after reading about his visit readers will encounter narwhals in different ways. If they come across a narwhal they might dwell longer, testing out or playing with this story, in hopefully new and critical ways. Who knows, they might wish to contemplate their 'angles of arrival', 'fragility and loss' as well as 'awe and awkwardness'. They might even be tempted to stare into a specimen's glass eye, and contemplate the subjectivities underwriting its seemingly 'natural' (that is, culture free) taxidermy forms. And they might look for different things when they encounter different natural history specimens and other scientific objects.

This points to pedagogical practices in which we share our personalised museum experiences with learners, whether in real-time or as preparation. Needless to say, this is quite a departure from the familiar museum worksheet. In more general terms, this analysis encourages learners and teachers to share their embodied feelings in science, in informal *and* formal settings.

At this point, we anticipate that some readers might be struggling with whether this museum visit should count as science education at all? After all, what science did the first author learn? How did he know that he learned anything? From a mainstream science education perspective, a response to these questions would likely take the form of a traditional 'learning outcome'. The very efficacy of this exhibit would rest on recounting a central linguistic narrative in one form or another, perhaps along the lines of "I have learned about Arctic ecosystems and their fragility to human actions". Such outcomes are important and we certainly do not discount them here. Indeed, in these terms, there are some limitations to this exhibit, including whether it is of sufficient depth or detail to convey complexities of Arctic biodiversity: the amount of text in the exhibit is limited largely to the aforementioned red warning sign.

However, in this chapter we have sought to extend what traditionally gets to count as science education. It has been an invitation to explore science education from a perspective of 'affect theory'. In contrast to familiar teaching and learning metaphors of *construction* we have turned to metaphors of *encountering* and *affecting*. In such terms, science and education are no longer solely equated with acquiring a particular language (in the form of a conventional learning outcome), but are also conceived as an active and ongoing embodied process of affecting and becoming affected in some ways. In such terms, science education is framed as an increasing ability to encounter an always affective/affecting scientific world.

The learner and the exhibit are now potentially transformed and their real-time, moment-by-moment lived, embodied, felt interactions seem much more important. We are now exploring how learners become part of the exhibit and how the exhibit therefore becomes part of the learner. Affect theorists would argue that all such encounters (whether in science or education) are co-constitutively shaped by our experiences and desires (Spinoza's 'affectus'). As pointed out earlier this is inclusive of our hopes, our fears, our joys and frailties—those anticipations, ambiguities, aspirations and hesitations that shape why, how and what we seek to achieve, or not. Learning science, in such terms, is not solely a conceptual outcome, but also an embodied condition. It is not simply about acquiring an answer, but it is also about embracing and realising continued beginnings.

A potential limitation of the exhibit in such terms is that during the time that the author was in attendance, it seemed that he was the only person immersed in the exhibit. For others, it appeared mainly an as 'object' in-passing, rather than a 'subject' of continuing exploration. This observation is worthy of a more extended analysis.

So What?

We are now left with a nagging question, "Why?"—Why should we seek to navigate the changing landscape of formal and informal science opportunities? What might we gain and lose by the arguments outlined in this chapter? There are unresolved political and ethical considerations that need to be reconciled here, too.

In concluding, let us return, for just one moment, to the encounter with the narwhal, guided by the question: What might be lost or gained within different encounters with narwhals? The narwhal exhibit certainly seems to have praiseworthy political intentions: seeking to convey a sense of biodiversity fragility and loss and the ways in which human actions are implicated. The underlying assumption is that if we know we will act: "Knowledge saves" (Haraway, 2008, p. 256). Although, like Donna Haraway, we long for representations of "multi-species flourishing" outside of the all too familiar "saving the endangered [fill in the blank]".

Notwithstanding such reservations, knowing about narwhals and the associated decline of Arctic ecosystems seems significant. Indeed, we suggest that school science curricula in Canada and around the world should focus more on research on Arctic flora and fauna. Much evidence suggests that it is undergoing dramatic changes. Yet our encounters with this narwhal allude to something else, something altogether more. They convey the importance of embodied, sensory affective interactions and relationships in the making. That education is a way of attending with, a way of feeling and relating to. Such discussions encourage considerations of subjective, emotional and moral values of our affectively charged bodily relationships in science education.

In *I and thou*, Martin Buber (1923/1958) offers a distinction in human existence mapped out in two modes of encountering. In the more common mode of 'I-it' encounters, experience is seen as a separate objectified object, or an 'it'. In the less common mode of 'I-thou', encounters rest on forging dynamic relationships. In this mode, through participation, both the 'I' and the 'thou' (the other) are transformed through their dynamically shifting rationalities. Buber suggests that this type of phenomenon is best described as a type of love, that is, love of an increasingly spiritual kind. Buber's thesis is that human life finds its meaningfulness in relationships, within increasing encounters of the 'I-thou' mode rather than the more common 'I-it' mode.

There is similar discussion within environmental education about the importance of forging more ethical and caring relationships with the organic world that we are part of (not separated from). We are thinking here, for instance, of Paul Shepard's (1982) and David Abram's (1996) common thesis that modern development of the self has compromised relationships with the organic world that surrounds us. As a consequence we have become blind to our destructive actions. At the very least, this perspective opens up questions of what relationships and subjectivities are we nurturing, and ought we be nurturing, within our science education encounters? We end this chapter with a question: How might science education learn to 'live better' with charismatic endangered creatures, such as narwhals, in the era widely being described as the Anthropocene? We suggest that affect theory might offer one response to navigating the changing landscapes of formal and informal science learning opportunities.

References

Abram, D. (1996). *The spell of the sensuous*. New York, NY: Vintage Books.

Ahmed, S. (2010). Happy objects. In M. Gregg & S. Seigworth (Eds.), *The affect theory reader* (pp. 29–51). London, UK: Duke University.

Alberti, S. (2010). *The afterlives of animals: A museum menagerie*. London, UK: University of Virgina.

Alsop, S. (2011). The body bites back. *Cultural Studies in Science Education, 6*(3), 611–623.

Barad, K. (2003). Posthumanist performativity: Toward an understanding of how matter comes to matter. *Journal of Women in Culture and Society, 28*(3), 801–831.

Barthes, R. (2005). *The neutral*. New York, NY: Columbia University.

Bennett, J. (2010). *Vibrant matter: A political ecology of things*. Durham, UK: Duke University.

Betelsen, L., & Murphie, A. (2010). An ethics of everyday infinities and powers: Felix Guattari on affect and the refrain. In M. Gregg & S. Seigworth (Eds.), *The affect theory reader* (pp. 138–161). London, UK: Duke University.

Buber, M. (1923/1958). *I and Thou* (R. Smith, Trans.). London, UK: Charles Scribner's Sons.

Deleuze, G. (1988). *Spinoza: Practical philosophy*. San Francisco, CA: Continuum.

Eisner, E. (2005). *Re-imagining schools: The selected works of Elliot W. Eisner*. New York, NY: Routledge.

Gregg, M., & Seigworth, S. (Eds.). (2010). *The affect theory reader*. London, UK: Duke University.

Haraway, D. (2008). *When species meet*. Minneapolis, MN: University of Minnesota.

Ihde, D. (1990). *Technology and the lifeworld*. Bloomington/Indianapolis, Indiana: Indiana University.

Jonas, H. (1982). *The phenomenon of life: Toward a philosophical biology*. Chicago, IL: University of Chicago.

Kirkman, R. (2007). A little knowledge is a dangerous thing: Human vulnerability in a changing climate. In S. Cataldi & W. Hamrick (Eds.), *Merleau-Ponty and environmental philosophy: Dwelling on the landscape of thought* (pp. 19–35). Albany, NY: State University of New York.

Latour, B. (1983). Give me a laboratory and I will raise the world. In K. Knorr Certina & M. Mulkay (Eds.), *Science observed: Perspectives on the social study of science* (pp. 141–170). London, UK: Sage.

Latour, B. (2005). *Reassembling the social: An introduction to A-N-T*. New York, NY: Oxford University.

Law, J. (2004). *After method: Mess in social science research*. New York, NY: Routledge.

Massumi, B. (2015). *The politics of affect*. Oxford, UK: Polity.

Payne, P., & Wattchow, B. (2009). Phenomenological deconstruction, slow pedagogy, and the corporeal turn in wild environmental/outdoor education. *Canadian Journal of Environmental Education, 14*, 15–32.

Poliquin, R. (2008). The matter and meaning of museum taxidermy. *Museum and Society, 6*(2), 123–134.

Poliquin, R. (2011). *The breathless zoo: Taxidermy and the cultures of longing*. University Park, PA: Pennsylvania State University.

Roehl, T. (2012). From witnessing to recording—material objects and the epistemic configuration of science classes. *Pedagogy, Culutre & Society, 20*(1), 49–70.

Rosing, J. (1999). *The unicorn and the Arctic Sea*. Newcastle, ON: Penumbra.

Shepard, P. (1982). *Nature and madness*. San Francisco, CA: Sierra.

Star, S. L., & Griesemer, J. (1989). Institutional ecology, 'translations' and boundary objects: Amateurs and professionals in Berkeley's Museum of Vertebrate Zoology, 1907–39. *Social Studies of Science, 19*, 387–420.

Staus, N., & Falk, J. (2013). The role of emotion in ecotourism experiences. In R. Ballantyne & J. Packer (Eds.), *The international handbook on ecotourism* (pp. 178–192). Cheltham, UK: Edward Elgar.

Stewart, K. (2010). Afterword: Worlding refrains. In S. Seigworth & M. Gregg (Eds.), *The affect theory reader* (pp. 339–355). London, UK: Duke University.

World Wildlife Fund [WWF]. (2015). *Unicorn of the sea*. Retrieved from https://www.worldwildlife.org/stories/unicorn-of-the-sea-narwhal-facts

Communicating Science

Susan M. Stocklmayer

Abstract Despite a growing understanding of the importance of inquiry learning and hands-on experiences in the classroom, science education generally operates within defined and somewhat old-fashioned curricula. There is much evidence that members of the general public, who are the products of this education, generally demonstrate an indifference to science and a lack of awareness of personal relevance of past or present research. Science education still has strong overtones of the deficit model, which has been rejected by science communication theorists and many practitioners for decades. The deficit model is fuelled by regular surveys of the public undertaken in Europe, the US and Australia which demonstrate a lack of knowledge of many simple scientific facts. In turn these surveys stimulate calls for improvement in general science literacy, which is seen to be deficient in most Western countries.

The principles of science communication place the needs of the 'audience' as the primary consideration, not the content of the scientific 'message'. This chapter examines how this might happen in the classroom, while retaining the requirement for formal science education to address the needs of a wide group of students which includes future scientists as well as those who will not continue with scientific study.

Keywords Science communication · Inquiry learning · Relevance

Is Science Education Providing for Lifelong Learning?

What are the important things to know about in science? If a student does not aspire to a scientific career, what aspects of a science curriculum will be relevant and useful? A brief look at the science confronting people today indicates that decisions are required every day from ordinary people about the food they eat, the environment in which they live, the insulation they select for their homes, the pesticide they put on

S. M. Stocklmayer (✉)
The Australian National University, Canberra, Australia
e-mail: sue.stocklmayer@anu.edu.au

© Springer International Publishing AG, part of Springer Nature 2018
D. Corrigan et al. (eds.), *Navigating the Changing Landscape of Formal and Informal Science Learning Opportunities*,
https://doi.org/10.1007/978-3-319-89761-5_5

their garden. They are also required to have knowledgeable opinions about the planet and its management. These are all science-based issues - but how well equipped are people to make these decisions?

Those concerned with communicating science to adults generally acknowledge that it is not necessary that people understand the finer details of climate science, genetic modification or nanotechnology. Rather, they need to know what the impacts of such developments are likely to be for them and the wider world. This implies that the science that is communicated to *students* must be presented in a memorable and relevant way, to form a sound basis for lifelong learning. This focus on more useful kinds of scientific information is well understood by those who conduct research into adult attitudes and knowledge, but is it common in a classroom context? Is it the general aim of informal learning provision, or science learning beyond the school gates?

In this chapter, some of the issues that concern the communication of science in the context of lifelong learning are reviewed and the challenges for teaching science across the science learning landscape (inclusive of formal and informal learning) are discussed. The idea of 'science literacy', used in many curricular statements and defined in many ways, is a powerful driver of science learning—but is it achievable? Perhaps science teaching and learning provision need to change to enable people to be a part of a democratic society. If so, can our knowledge about public engagement enable a reframing of how we approach teaching and learning? How should the providers of science experiences approach this challenge and what should be the goals of communication in the science learning landscape?

Science and the Public—A Gap in Understanding?

There is no doubt that communicating science is difficult. Cultivating the art of 'translating' the science for those less well versed in its complexities and jargon has led, over several centuries, to the rise of a group now known as 'science communicators'. They number among their ranks Galileo Galilei (1564–1642), Michael Faraday (1791–1867) and Lawrence Bragg (1891–1971). Galileo was one of the first to communicate in the language of non-scientists, in his case Italian; Michael Faraday enchanted lay audiences with iconic demonstrations and simplified explanations; and both he and Bragg wrote hints on how to do this well.

Science communication as a formal discipline, however, is of much more recent origin. Formal frameworks around which research into science communication is conducted have, generally, only been established since the 1980s. Science journalism as a profession is part of this framework, but by no means all of it. In the 1990s, many tertiary programmes in science communication were developed, but the discipline has changed substantially since that time in the light of new perceptions of what it means to communicate science.

Science communication research has focused on the relationship between scientists and the public. However, this simple statement hides a multitude of complexities and nuances that, particularly since 2000, have emerged as issues affecting the communication of science. There are, for example, many 'publics', who range from the well-informed to the uninterested, from the 'don't cares' to the techno-geeks, from politicians to farmers or the neighbours next door. Navigating this complexity and understanding how better to engage diverse individuals and communities, especially across cultures, is a task that requires inspired research and deep understanding from those within the discipline.

In the beginning, the task seemed more straightforward than it does today. From the 1950s, it was thought that the public needed to be educated about science because "science was seen as a key to national prosperity and the need for the public's support for science was crucial" (DeBoer, 2000, p. 585). For the first time, the general public's knowledge of science was called into question. By the 1980s, a movement known as the 'Public Understanding of Science' (PUS) had become concerned for public 'scientific literacy'.[1] The Royal Society of London (1985) articulated goals for the public that were, in summary, a need to understand science in order to: foster economic prosperity; comprehend science in everyday life; be able to participate in democratic issues; appreciate science as culture; and appreciate how and where funding for research was being spent. Efforts were made to improve public education in science. The assumption was that increased knowledge of science would result in increased acceptance of science. The PUS movement prompted the rise of science centres, festivals and science events, all aimed at improving a loosely defined 'scientific literacy' by informing an uninformed public.

According to Paisley (1998), scientific literacy in the USA had its origins in the post-Sputnik programmes, undertaken not only "to move students toward scientific careers" but also "to engender public support for the costs and risks of Cold War science" (p. 70). Thus, the major impetus for improving public understanding was unequivocally economic on both sides of the Atlantic and was firmly grounded in a push for an international edge over industrial competitors. It also formed the core of science communication research.

The notion of scientific literacy became strongly linked to public knowledge of science. Unsurprisingly, therefore, calls for the enhancement of public understanding resulted in a need for baseline measurement of that understanding (Bodmer, 1985). The resultant public surveys, in particular the one described by Durant, Evans, and Thomas (1989), indicated that the public/s had scant knowledge of science. This became an issue of international concern. In the Durant et al. survey, large samples of the public were interviewed in the UK and USA. Although the survey also examined levels of interest in science and understanding of scientific methods, its measurement of knowledge was stated to be the most important element. Despite the fact that "there was considerable tacit understanding of the processes of scientific inquiry" (p. 12) it was the knowledge part of the survey that attracted the most attention and has continued to do so ever since (Bauer, 2009). The knowledge section consisted of 22 factual items requiring the responses of 'agree',

---

[1] This term appears to have surfaced around 1958, but remained largely undefined until the 1970s (DeBoer, 2000, pp. 587–588).

'disagree' or 'not sure'. The survey, according to the authors, was of an "extremely elementary nature" with statements such as "hot air rises" or "diamonds are made of carbon". Overall, however, the public did not do well.

> The authors concluded that the public was ill equipped to deal with increasing influences of science and technology, either in people's daily lives or in terms of changing workplace demands. Gloomily, it was concluded that prospects for democratic processes were poor, in terms of public debate and decision-making. (Stocklmayer & Bryant, 2012, p. 3)

When socio-demographic factors were factored in, results became somewhat worse: older people, females and 'working-class people' on average scored less well. "It is regrettably a commonplace that in our culture women tend to be less interested in and involved in science than men" (Durant et al., 1989, p. 14).

From the result of such surveys, and perceptions that the public lacked the necessary background to understand and engage with scientific endeavours, the term 'deficit model' was coined (Layton, Jenkins, McGill, & Davey, 1993; Wynne, 1991; Ziman, 1991). The fundamental assumptions of the 'deficit model' were that members of the general public knew and understood very little science. This basic knowledge that the public lacks is knowledge that scientists possess. It is important that the public should have it too and therefore lay people require further education.

Initiatives to provide for such public education gave rise to the PUS movement, which throughout the 1990s dominated science communication in theory and practice. In this same decade, questions from the Durant et al. (1989) survey (now known as 'The Oxford Scale') were incorporated into national and international measures of 'scientific literacy' or, sometimes, 'civic scientific literacy' in the US (Miller, 1998). The US National Science Foundation's 'Science and Engineering Indicators' and, in Europe, the Eurobarometer measures have used questions from this survey right up to the present.

> The items are intended to capture one or more dimensions of what Miller (1998) refers to as 'civic scientific literacy'…. Most of the true/false items tap what might be termed 'textbook' type knowledge across a range of scientific domains. (Allum, Sturgis, Tabourazi, & Brunton-Smith, 2008, p. 38)

This textbook knowledge is represented by statements such as "The oxygen we breathe comes from plants" and "It is the father's gene that decides whether the baby is a boy or a girl". Both these statements are problematic in the way they are written and even may be deemed incorrect. Nevertheless, the first is often used to conclude that people do not understand photosynthesis and the second, basic genetics. Both have been repeated endlessly, over decades, in the international indicators mentioned above.

The survey has been used in many countries to compare a range of demographic features. For example, a consistent gender difference has been observed. It is notable, however, that on questions related to biological science, which are in the minority in the survey, women score higher than men. Nevertheless, the finding has been unconditionally accepted: "… given the fact that women tend to be less knowledgeable about science than men … it might be fruitful to explore the underlying factors" (Sturgis & Allum, 2001, p. 429). Countries, too, have been compared.

Despite many apologists for the survey asserting that it should be seen only as an overall *indicator* of attitudes, engagement and knowledge (National Science Foundation, 2014; Sturgis & Allum, 2004), country averages have been published and compared using common questions that frequently include the two mentioned above.

Stocklmayer and Bryant (2012) have pointed out that scientists themselves do not always get the answers 'right' and that many questions are open to several interpretations. As a measure of knowledge, these surveys—and others like them—are fundamentally flawed. They have been criticised since the early 1990s as measuring rote learning and memory, not deep understanding. Statements such as "the Earth moves around the Sun", for which responses of approximately 70% agreement have not shifted around the world since 1989 (National Science Foundation, 2014), are now often seen as futile. In an early comment, Fayard (1992) said,

> Let's stop persecuting people just because they don't think like Galileo! I confess that I myself have never woken up in the morning saying 'the movement of the Earth on its axis is such that the Sun can be seen in the east'—in my daily life the Sun moves round the Earth. (p. 15)

Thus there has been a considerable emphasis on knowledge of facts that, over time, has scarcely altered. From 1998 to 2012 in the US, for example, the question about the father's gene had a response of 'true' ranging from 61% to 66%. It was highest in 1999. Similarly, "the universe began with a huge explosion" had the response of 'true' between 32% and 39% for all years except 1988, when it was higher. Some questions have shown small gains, the most notable being a steady rise from 25% to 51% for "antibiotics kill viruses as well as bacteria" but the stubbornly unchanging percentages for most questions over 25 years must, surely, indicate that these findings are somewhat pointless because no action can be taken to address them.

Despite the criticisms, the survey has formed the basis of measurement of public knowledge in Asia, Europe and the US since its inception and, more recently, also in Australia. One is driven to conclude that the reason for the repetitious administration of these questions is because they are able to measure *something*, rather than that the survey is intrinsically valuable. Further, the information that, for example, many of the public do not know that the Earth orbits the Sun is useful to lobby for more funding for science education. It has considerable shock value. This is not new:

> The importance of science in everyday life is often stressed... At the same time, it is also stressed that the 'man in the street' has little conception of what science is and how it advances. Scientists are often the loudest in proclaiming this popular ignorance, especially when they want to get money to support their schemes. (Bragg, quoted in Porter & Friday, 1974, p. 1)

In summary, in the 1980s the movement known as the Public Understanding of Science (PUS) became concerned for public (civic) scientific literacy. Efforts were made to improve public education in science, assuming a deficit in public knowledge that needed to be filled. The drivers for this movement were, in the main, those articulated by the Royal Society in 1985, especially the economic imperative. In the 1990s, therefore, tertiary programmes in science communication were developed based on a foundation of 'getting the message across' and 'knowledge transfer'. Even as these ideas were being researched and practiced, however, concurrent events were about to cast the ideas of one-way transmission of knowledge into complete disarray.

Does Knowledge of Science Facts Really Matter?

During the 1990s, a number of events in Europe caused the efficacy of the PUS movement to be questioned—disasters such as foot-and-mouth disease, genetically modified crop problems, and BSE ('mad cow disease') provoked a crisis of trust in science. In the UK, the result was a parliamentary commission of inquiry, which spent a year investigating the relationship between science and society. The resultant findings, known colloquially as the 'House of Lords Report' (House of Lords, 2000) were, in summary, that the PUS movement had been ineffective in engaging with the uninterested public and was top down and arrogant. It was noted that the arguments put forward in 1985 for PUS by the Royal Society placed responsibility for acquiring such knowledge on the public: "they need to learn more science because......". This approach, it was now agreed, was both unreasonable and unrealistic. Instead, a new science–society relationship had to be formed on a basis of respect and interaction, and a new term devised to replace 'PUS'. The previous model of communication, which was essentially one-way, was to be re-visioned. The deficit model was not appropriate. As I have previously pointed out:

> The one-way model has also been widely criticised for its underlying implication that the transmission of information is from 'expert' to 'layperson'—implying that the public is somehow deficient in their understanding of science. This model has now been comprehensively rejected in favour of a style of engagement that respects public knowledge as well as the knowledge of scientists, and regards the public and scientists as equal players in science communication endeavours... Terms used to describe this more equal relationship include public engagement, dialogue, knowledge sharing and knowledge building. (Stocklmayer, 2013, p. 20)

The need for change came as a considerable shock to many in the scientific community who had embraced the PUS movement. Correspondence across the emails at that time had overtones of disillusion and disappointment. For those who had devoted much time and energy into giving public lectures, setting up festival events and so on, the news that this had appealed only to the converted or the semi-converted was disheartening. Nevertheless, there were many researchers who had been suggesting for more than a decade that communicating science was more complex than simple transmission (e.g., Wynne, 1991, 1992; Ziman, 1991). There was therefore an existing body of research that was useful in framing a new approach.

The ideal mode of communication of science has shifted from one-way transmission to some form of two-way, participatory practice that incorporates dialogue and consensus, decision-making and policy formulation (see, for example, Pedretti & Navas-Iannini in their Chapter "Pregnant Pauses: Science Museums, Schools and a

Controversial Exhibition"). The basis for engagement is the premise that respectful interaction between scientists and the public will result in the sharing of knowledge and, perhaps, construction of new knowledge. In this framework, 'the public' represents whoever in the non-scientific community is concerned with the issue at hand. There remains confusion, however, over one-way communication, which according to several authors *always* implies a deficit model. Some claim that the deficit model has never really gone away (Bauer, Allum, & Miller, 2007; Wynne, 2006). Trench (2006) stated that "a deficit model remains the default position of scientists in their public activities and underpins much of what is proposed by public officials in their promotion of science" (p. 1). Wynne (2006) reviewed the history of views about the public's trust in science and concluded that the deficit model is as pervasive as ever. Kim (2007) argued that "this is the dominant communication strategy and the reigning behavioural theory" (p. 288).

From the perspective of science communicators, however, with a range of outlets for their communication such as the media and live science presentations. it is impossible to ignore one-way communication strategies. What is different is that the underlying *intent* has changed (Stocklmayer, 2013). Thus,

> ... examples such as the inclusion of informative articles in the press, screening a television documentary, placing science on the Internet or presenting a new exhibition in a science centre ... are certainly overwhelmingly one-way in their design and, therefore, intent. There is clearly no expectation by the writers, designers and producers that they will engage in two-way communication, but rather that they are 'transmitting' information to whatever audience is willing to listen, play, read or watch. All these examples nevertheless contribute to a view of scientific knowledge as knowledge worth having, interesting or important to a variety of people. (Stocklmayer, 2013, p. 22)

The foundation for this kind of communication must be mutual respect—but, unfortunately, respect for the knowledge of the 'lay public' somehow got lost in the years of the deficit model when 'educating the masses' was the goal. It had, however, always been part of the mindset of the great communicators from earliest times: "[The lecturer's] whole behaviour should evince a respect for his audience, and he should in no case forget that he is in their presence... (Faraday, quoted in Porter & Friday, 1974, p. 8).

Is the Public Generally Indifferent to Science?

There have been several surveys since 2000 seeking to assess public attitudes to science and they overwhelmingly indicate that the public generally supports scientific enterprise (e.g., Castell et al., 2014; Lamberts, Grant, & Martin, 2010; National Science Foundation, 2014; Searle, 2014). People do not, however, feel well informed about science and would like to know more. (It does not, of course, follow that 'feeling informed' necessarily means being correct in one's knowledge. If views are already polarised, then further information may simply reinforce those views.) In addition, a gender gap remains:

Women are less likely than men to feel informed about science and often feel less confident in engaging with it. When it comes to studying and working in science and engineering, women tend to be less positive. This gender divide may develop before adulthood, with far fewer young women than young men participating in science or engineering clubs at school. At the same time, it should be acknowledged that women appear to play a particularly important role in informal science learning. People are more likely to go with their mother rather than their father to science-related leisure or cultural activities, and women themselves are more likely to take others with them rather than going alone. (Castell et al., 2014, p. 6)

Although these remarks refer specifically to the UK, they are generally applicable in other Western countries. There may also be a gender difference in the way that people retrospectively regard school science. According to the Ipsos MORI survey,[2] about 17% of men thought that school had "put them off" science compared with 30% of women (Castell et al., 2014, p. 108). School science was thought to be useful in daily life by about half the sample, although this finding also had embedded gender differences.

Although most surveys indicate a strong correlation between support for science and levels of education, it is apparent that important elements such as the way science works, or the idea that science is, essentially, a way of thinking rather than a body of factual content, are not well understood. In general, the conclusion may be drawn from these surveys that overall, in most countries, there is high interest in science and a desire for more information. People would like more communication from scientists themselves and would like the public to be consulted about scientific issues early in the implementation process. There is, however, considerable misunderstanding about the way science research is conducted, peer reviewed and implemented. Risk and uncertainty are seen as confusing.

It seems clear from these findings that school science has not equipped the public to address issues of concern. It is especially ineffective for women. Generally, the focus on factual content has diminished the importance of critical thinking and an understanding of the processes of science. When the history of science education itself is reviewed, however, it is evident that teaching these things was originally an educational goal, but that this has changed over the intervening hundred years.

What Is the Point of Science Education?

Looking at the history of science education in the UK and the US, we can distinguish slightly different original aims that, over time, have merged into common perceptions of what school science should be aiming to do. School science was firmly established in the UK by the time Bragg was born in 1891, with a general goal of educating a somewhat elite group of male students to become scientists (Layton, 1993). Those who urged the need for inclusion of science education in the

---

[2] **Ipsos MORI** is the second largest market research organisation in the United Kingdom, formed by a merger of **Ipsos UK** and **MORI**, two of Britain's leading survey companies.

curriculum included such luminaries as Thomas Huxley, Herbert Spencer, Charles Lyell, Michael Faraday, and John Tyndall (DeBoer, 2000). Higher goals for school science were a fundamental part of their argument:

> The humanities were firmly entrenched as the subjects that were thought to lead to the most noble and worthy educational outcomes. Scientists had to be careful when arguing the utility of science not to present science as too crassly materialistic and without higher virtue. So in addition to discussing the practical importance of science in a world that was becoming dominated by science and technology, they also said that science provided intellectual training at the highest level—not the deductive logic that characterized most of formal education, but the inductive process of observing the natural world and drawing conclusions from it. (DeBoer, 2000, p. 583)

Even as these high goals came to dominate science education, however, there were calls for a more relevant science that would equip the general population to be critical thinkers who recognised the place of science in everyday life:

> During the early years of the 20th Century, largely because of the influence of writers such as Dewey, science education, and education in general, was justified more and more on the basis of its relevance to contemporary life and its contribution to a shared understanding of the world on the part of all members of society. (DeBoer, 2000, p. 583)

Unfortunately this idealism did not last and, by the 1930s, curriculum designers were being criticised for making science too relevant, with insufficient focus on understanding of scientific principles. Science, it was felt, should not only be taught in terms of its usefulness to the individual and to enable participation in a democratic society, but as "a powerful cultural force and a search for truth and beauty in the world" (DeBoer, 2000, p. 584).

In the US, a 'Committee of Ten' had been appointed in 1892 to decide on a curriculum, which included science (National Education Association of the United States, 1894). The emphasis was, at the outset, more upon citizen science than preparation for college entrance, but by the 1950s science was in the spotlight because of perceived lagging behind in the Space Race. In a retrospective review of the Sputnik years, Rutherford (1997) asked: "Should, progressive, child-centered education or basic, discipline-centered education have precedence in the schools?" and "Should priority be given to building the nation's scientific capability or to creating nationwide science literacy?" (p. 3).

Despite subsequent introductions of alternative curricula that focused on 'science and society' (but were often considered to be of lesser value than 'pure science' curricula) the 1980s saw a resurgence of the emphasis on elitism in science in school. By the 1990s, Science–Technology–Society (STS) studies tended to be abandoned in favor of a formal separation of science and technology. New courses in science were more conceptually difficult than before and were, once again, overtly designed for the more capable student. This assumption of elitism is often still reported: the ASPIRES study (Archer, Osborne, & DeWitt, 2012) found that students and their parents still consider school science to be relatively difficult, and that only the 'smart kids' can do science. Further, there are educational systems that weight school leaving scores in favour of science topics, particularly the physical sciences. This elite approach was accompanied in Europe and the United States by continued rising concerns for the level of science literacy in the general population, discussed earlier in this chapter.

The rationale for scientific literacy in the US *Next Generation Science Standards* (NGSS Lead States, 2013) has a strong emphasis on economic purposes, with connections to a perceived reduction of the nation's competitive edge. In addition, it states:

> Beyond the concern of employability looms the larger question of what it takes to thrive in today's society. Citizens now face problems from pandemics to energy shortages whose solutions require all the scientific and technological genius we can muster. Americans are being forced to increasingly make decisions—including on health care and retirement planning—where literacy in science and mathematics is a real advantage.

These sentiments are echoed in the *Science and Society Programme* in the UK, which has a vision "that all citizens share in the development and contribution of science to UK culture, quality of life, sustainable economic development and growth, and feel a sense of ownership about the direction of science and technology" (Castell et al., 2014, p. 9).

Is Scientific Literacy Ever Attainable?

The desirability of 'scientific literacy' has thus, over several decades, been strongly linked to the arguments put forward by the Royal Society of London, mentioned earlier. It has implications for the economy, for support for science, for democracy and culture. There have, therefore, been various attempts to define the term. Some focus on facts and concepts while others emphasise more subtle issues related to science and society.

Laugksch (2000) conducted an overview of "over 330 journal articles, conference papers, project descriptions, project reports, and editorials related to scientific literacy [that] were found to have appeared in the literature between 1974 and 1990... with the vast majority being published after 1980" (p. 73). He identified four interested groups for whom a workable definition of scientific literacy would be important. Each of these groups focuses on a different but related audience. The group concerned with formal education has, not surprisingly, tended to focus on the definitions relating to a scientifically literate school student that appear in many formal curricula.

The remaining three groups are concerned with members of the general (lay) public. Of these, the first group "is essentially concerned about the extent of the general public's support for science and technology, as well as the public's participation in science and technology policy activities" (Laugksch, 2000, p. 75). For this group, it is important to know more about an individual's knowledge base and their sources of scientific and technical information, together with attitudes to science and science policy. Definitions are, accordingly, framed in these terms. The second group "are concerned with the construction of authority with respect to science (i.e., organisational forms of ownership and control of science)" (p. 75). For this group, understanding how people access and use science in everyday life and apply science in a personal context is important. The third group constitutes the informal science community, which consists of professionals who include

> ... relevant personnel involved in science museums and science centers, botanical gardens and zoos, as well as members of creative teams involved in science exhibitions and science displays. Science journalists and writers, and relevant personnel involved in science radio programs and television shows complete this interest group. (p. 75)

In terms of formal education (the first group), the many and various definitions of scientific literacy that have been proposed are not especially relevant to this chapter. A particularly broad and workable definition, however, was published in Australia. In the *National review of the status and quality of teaching and learning of science in Australian schools,* Rennie, Goodrum, and Hackling (2001) argued that the broad purpose of teaching science in the compulsory years of schooling is to develop scientific literacy for all students:

> Scientific literacy is a high priority for all citizens, helping them
>
> • to be interested in, and understand the world around them,
> • to engage in the discourses of and about science,
> • to be sceptical and questioning of claims made by others about scientific matters,
> • to be able to identify questions, investigate and draw evidence-based conclusions, and
> • to make informed decisions about the environment and their own health and well- being. (p. 7)

Definitions concerning the general public put forward by the remaining three groups have tended to focus around knowledge. Some have had particularly high aspirations. In Miller's (1998) analysis of civic scientific literacy, for example, he states that a score of 67% or above on the knowledge section of the Durant et al. (1989) survey indicates that the respondent is well informed. If respondents gained a similar score on a test about scientific inquiry, they were deemed 'civic scientifically literate'. One without the other meant that the participant was "partially civic scientifically literate" (p. 216). From 2010 data we may conclude that "adult men in the USA are currently only partially literate, while women fall below the line (National Science Board, 2010, Table 7.8). In 1995, 12% of Americans were deemed fully literate, while Europe recorded a score of 5% (Stocklmayer & Bryant, 2012). Miller (2010) also suggested that scientific literacy might be measured in terms of a level of understanding sufficient to read science and technology stories written at the level of the *New York Times* 'Tuesday Science' section, although analysis of reading level indicates that this measure would require a reading ability equivalent to the very top of high school, or even early tertiary study.

Three aspects of science literacy were identified by Shen (1975) and summarised by Rennie and Williams (2002) as follows:

> Practical scientific literacy is having the kind of scientific knowledge that can be used to solve practical problems. Civic scientific literacy enables citizens to be aware of science and science-related issues and to think about and make decisions in a democratic process. Cultural scientific literacy is knowing something about science as a major human achievement. (p. 708)

The Organisation for Economic Co-operation and Development (OECD)'s PISA (2015) statement embraces a wider view:

As individuals, we make decisions and choices that influence the directions of new technologies, e.g., to drive smaller, more fuel-efficient cars. The scientifically literate individual should therefore be able to make more informed choices. They should also be able to recognise that, whilst science and technology are often a source of solutions, paradoxically, they can also be seen as a source of risk, generating new problems which, in turn, may require science and technology to resolve. Therefore, individuals need to be able to consider the implications of the application of scientific knowledge and the issues it might pose for themselves or the wider society. Scientific literacy also requires not just knowledge of the concepts and theories of science but also a knowledge of the common procedures and practices associated with scientific enquiry and how these enable science to advance.... (pp. 3–4)

In short, the OECD's PISA Framework defines scientific literacy as "the ability to engage with science-related issues, and with the ideas of science, as a reflective citizen" (p. 7).

Ogawa (2013) has said, however, that the goals of science literacy may not be practical:

Should we continue to endeavour to achieve an ideal ultimate goal where all of the citizens hold a certain level (satisfactory to the scientist community) of scientific literacy and/or engagement—that is, an ideal future community with individuals with perfect scientific literacy and perfect engagement? While this indeed may be an ideal state, unfortunately we cannot overlook the fact that the diversity of science literacy levels among the community has remained rather stable or unchanged despite various intentional remedial efforts. This ideal is, therefore, really "ideal". It currently serves as the ultimate goal for certain groups, but these groups also need to accept a reality in which the diversity in community levels of 'scientific literacy' is not diminishing. These groups may need the wisdom to set up more "practical" goals along the way. (pp. 11–12)

It seems, therefore, that the whole notion of scientific literacy as an educational goal is ill-conceived and, probably, unattainable.

Can Science Really Engage the Public?

The term 'engagement' has connotations of positive action on the part of members of the public, whether they engage interactively, reactively or proactively. So what does it mean to 'engage' with an issue, or to 'engage' with scientific ideas? It seems that the word has a variety of meanings, from participation in science-based activities to debate and discussion with scientists or other members of the public. For example, Poliakoff and Webb (2007) have defined public engagement in science as "any scientific communication that engages an audience outside of academia" (p. 244). Others have specified outcomes of policy-making or agenda-setting through a range of mechanisms that include 'dialogue events', cafés scientifique, focus groups, and so on. "However, few academics and governments attempting to 'engage in engagement' are clear about their goals and desired outcomes, and whether or not the processes they facilitate are likely to meet these ends" (Powell & Colin, 2008, p. 127). These authors suggest that both citizens *and* scientists need training in how to engage with each other and that far greater support is needed at the institutional level for 'engagement events'. They also question whether it is realistically possible for groups drawn from the general public to have real influence in the scientific and politico-scientific domain.

In 2010, the Australian Government launched *Inspiring Australia, a national strategy for engaging with the sciences*. This landmark document stated:

> The aspirational goal is for a scientifically engaged Australia—a society that is inspired by and values scientific endeavour, that attracts increasing national and international interest in its science, that critically engages with key scientific issues and that encourages young people to pursue scientific studies and careers. (Australian Government, 2010, p. xiii)

Once again, the way in which this engagement is to occur was not well defined, although

> [i]n order to achieve a scientifically engaged Australia, it will be necessary to develop a culture where the sciences are recognised as relevant to everyday life and where the government, business, and academic and public institutions work together with the sciences to provide a coherent approach to communicating science and its benefits. (Australian Government, 2010, p. xiv)

In this regard, 15 recommendations for action were implemented, most incorporating some form of outreach to the community or involvement in activities such as citizen science. Familiar outreach of this kind often involves one-way communication for which people attend an event such as a lecture or a festival presentation or visit a venue such as a science centre or zoo. As has been discussed above, these people are most likely to be those with a pre-existing interest in science. Research carried out at these venues for informal learning therefore gives hints on how to 'engage' the more committed visitor, but not the uninterested or unengaged.

The key to successful engagement, however defined and no matter what the mode of communication, is relevance. It has been known for decades, through constructivist theory, that learners build knowledge on what they already know. However, this idea is more complex that it first appears because although a person may already know something, they may not use it successfully to extend their understanding unless it is *personally* relevant to them. This presents a challenge for communicators if they are unaware of personal frameworks and how to map into them. The initial 'hook' to interest and, therefore, to engage is critical.

In an extended analysis of what stimulates engagement, Walker (2012) mentions that interest depends on two things: the novelty of the topic and a person's ability to understand it (Silvia, 2006). Interest is therefore closely linked to a person's past experiences and the links they can make to the topic at hand (Stocklmayer & Gilbert, 2002). Interest is also closely linked to affective elements (e.g., Izard 2007) and fostering enjoyment (e.g., Power & Dalgleish, 2008). Interest and enjoyment are stimulated by surprise and curiosity, and the outcome is heightened relevance (Walker, 2012).

A surprising event or piece of information will be followed by asking 'Why?' In a study of over 400 'discrepant events' in the classroom, Liem (1987) found that once students became curious, the drive to learn more on their own was very powerful. Curiosity is essentially driven by being conscious of an information gap in one's knowledge—a need to know. It is different from interest in subtle ways, since one can be interested without necessarily seeking to fill a knowledge gap. Informal learning providers have, to some extent, already embraced these ideas of personal relevance and should consciously seek to highlight personal interest and enjoyment in communicating science.

Are Informal Learning Activities Making Matters Worse?

What, therefore, are the important things to know about in science? From a science communication perspective, a brief look at the science confronting people today indicates that chemistry, physics and other sciences need to be angled towards real-world issues. The emphasis on rote learning, so long recognised as non-memorable and not especially useful, must change to enable people to be a part of a democratic society for whom science is so often part of decision-making. In the Ipsos MORI survey (Castell et al., 2014), topics raised with the public for consideration of interest and engagement included agri-science, food security and emerging energy technologies such as wind farms, fracking and carbon capture and storage. The challenge for educators is to enable people to feel confident in discussing such issues.

It is clear that the formal curriculum is not about to change to any marked degree, to be more relevant to later life and to enable the kinds of discussion and debate that are seen as civic responsibilities. It is well known historically that, no matter what cosmetic changes may be applied to curricula, the tendency to revert to what was previously done in the classroom is very strong. The drivers of performance, such as the PISA test, hold great power to determine what should be learned, and in what framework, and teachers of science are greatly influenced by the implied judgement—and perhaps threat—of the test. Tests such as PISA thus become the actual target for teaching and learning science in the view of many schools and many educational systems. In 2014, however, almost 100 senior academics, in an open letter to the OECD published in *The Guardian* newspaper, registered their concern about the educational consequences of this kind of science testing across nations:

> By emphasising a narrow range of measurable aspects of education, PISA takes attention away from the less measurable or immeasurable educational objectives like physical, moral, civic and artistic development, thereby dangerously narrowing our collective imagination regarding what education is and ought to be about. (*The Guardian* online, 2014)

They suggested that, in determining what PISA should really be about, the OECD should

> Make room for participation by the full range of relevant constituents and scholarship: to date, the groups with greatest influence on what and how international learning is assessed are psychometricians, statisticians, and economists. They certainly deserve a seat at the

table, but so do many other groups: parents, educators, administrators, community leaders, students, as well as scholars from disciplines like anthropology, sociology, history, philosophy, linguistics, as well as the arts and humanities. (*The Guardian* online, 2014)

Providers of informal learning have a responsibility to move away from these drivers of formal curricula. However, wherever informal learning is regarded chiefly as a bridge into the classroom—a bridge which helps school science to be better understood—it is reinforcing the view that this science, this body of knowledge, is the one that is most valued. It is, therefore, clearly difficult for many informal learning providers to depart from the curricular view, especially when science centre exhibitions and popular science events are designed with school group visitors in mind. The teachers need to believe that the visit will be valuable back in the classroom, and are not likely to come on an excursion if that value is not perceived. Thus the science curriculum is reinforced informally and the cycle continues.

Science communicators engaging with a public outside of school science, however, know that the science that is important is often quite different from that which is formally learned. The 'Big Ideas' of the curriculum should certainly be addressed, but the fine detail composed of endless facts, including jargon-ridden terminology and labelling, needs to be revised. It seems to me amazing that, after so many decades of recording muddles and misconceptions, of noting that scientific concepts are poorly understood and not well remembered, and of complaining about a general public who are deemed scientifically illiterate, we nevertheless continue to perpetrate a vision of science education simply as knowledge of the domain.

The knowledge we expect students to have by the end of middle/high school encompasses information and concepts from all the major scientific disciplines, to a depth that students cannot possibly embrace. The teachers tasked with imparting this knowledge are supposed to be able to deliver that depth, with understanding not just of concepts but of a wide range of applications including information about the latest research and, if possible, strategies using a variety of new technologies to help in finding out yet more facts. This simply cannot be achieved within current educational frameworks and, even if it were achievable, the students would soon forget whatever information is not reinforced in later life. All the evidence from educational research says that they do.

The aspirational goal for universal scientific literacy is too high. So what should those responsible for informal learning do? It seems to me that providers of informal learning opportunities to school-age students have a responsibility to step outside the narrow school curriculum to enable learning on a wider, more relevant scale. They should focus on attitudinal outcomes, with emphasis on individual learners (as exemplified by Steve Alsop in his Chapter "Encounters with a Narwhal: Revitalising Science Education's Capacity to Affect and Be Affected"). This can only occur if there is an understanding of what will be important in later life. It is clear that the conventional curriculum cannot be ignored, but surely the providers of informal learning opportunities have more to do than, say, run workshops on augmenting practical opportunities within the confines of the formal curriculum, or provide informal contexts to enable more effective learning of otherwise boring, complex and irrelevant science.

Many involved in informal learning see an important role in narrowing the learning gap created by those teachers who are 'out of their comfort zone'. The need for someone to interest students and help the teachers caught in this educational trap is pressing. At the same time, however, those who fill this need should try to address the wider picture discussed above. It is tempting, indeed, to see such activities as useful and important, but the world of informal learning can offer so much more.

It is not enough to provide an educational lifeline for survival of the school curriculum. We should speak out about its futility and its impossible aims and, if we truly desire 'scientific literacy', we should begin to think about what this really means. Especially, we should consider the aims for individuals, for particular communities, and for the nation.

Science communicators themselves therefore need to have clear goals for the outcomes they wish to achieve. Ogawa (2013) has said that science communicators, together with those with whom they communicate, should "set *in advance* their ideal goals for a particular community in terms of science communication" (p. 8). This implies a clear vision for the relationship between science and society for that community. In this regard, Rennie and Stocklmayer (2003) said that it is important that people feel that science and technology lie within their own interest and their personal lives, that they have ownership of their nation's science, that they can understand the impact of new technologies, that they can access and use scientific information, and that they should believe that the scientific community respects and values their knowledge and concerns. All of these are aspirational aspects of engagement that will prove hard to realise. Little of the school curriculum addresses these goals. People have said for many years and in many different ways that making connections with conventional science is often too difficult for them, but that relevant science is interesting and important. It is the role of science communicators to discover how to achieve this relevance and to give a new dimension to the term 'informal learning'.

References

Allum, N., Sturgis, P., Tabourazi, D., & Brunton-Smith, I. (2008). Science knowledge and attitudes across cultures: A meta-analysis. *Public Understanding of Science, 17*, 35–54.

Archer, L., Osborne, J., & DeWitt, J. (2012). *Ten science facts & fictions: The case for early education about STEM careers*. London, UK: King's College London.

Australian Government. (2010). *Inspiring Australia—A national strategy for engagement with the sciences*. Canberra, Australia: Questacon.

Bauer, M. W. (2009). The evolution of public understanding of science discourse and comparative evidence. *Science Technology and Society, 14*, 221–240.

Bauer, M. W., Allum, N., & Miller, S. (2007). What can we learn from 25 years of PUS survey research? Liberating and expanding the agenda. *Public Understanding of Science, 16*, 79–96.

Bodmer, W. F. (1985). *The public understanding of science*. Report of a Royal Society *ad hoc* Group endorsed by the Council of the Royal Society. London, UK: Royal Society.

Castell, S., Charlton, A., Clemence, M., Pettigre, N., Pope, S., Quigley, A., ... Silman, T. (2014). *Public attitudes to science 2014*. Retrieved from https://www.ipsos-mori.com/Assets/Docs/Polls/pas-2014-main-report.pdf

DeBoer, G. E. (2000). Scientifc literacy: Another look at its historical and contemporary meanings and its relationship to science education reform. *Journal of Research in Science Teaching, 37,* 582–601.

Durant, J. R., Evans, G. A., & Thomas, G. P. (1989). The public understanding of science. *Nature, 340,* 11–14.

Fayard, P. (1992). Let's stop persecuting people who don't think like Galileo! *Public Understanding of Science, 1,* 15–16.

House of Lords. (2000). *Third report of the select committee on science and society.* London, UK: House of Lords. Retrieved from http://www.publications.parliament.uk/pa/ld199900/ldselect/ldsctech/38/3801.htm

Izard, C. E. (2007). Basic emotions, natural kinds, emotion schemas, and a new paradigm. *Perspectives on Psychological Science, 2*(3), 260–280.

Kim, H. (2007). PEP/IS: A new model for communicative effectiveness of science. *Science Communication, 28,* 287–313.

Lamberts, R., Grant, W. G., & Martin, A. (2010). *Public opinion about science.* Canberra, Australia: The Australian National University.

Laugksch, R. C. (2000). Scientific literacy: A conceptual overview. *Science Education, 84*(1), 71–94.

Layton, D. (1993). *Technology's challenge to science education.* Bristol, UK: Taylor and Francis.

Layton, D., Jenkins, E., McGill, S., & Davey, A. (1993). *Inarticulate science? Perspectives on the public understanding of science and some implications for science education.* East Yorkshire, UK: Studies in Education.

Liem, T. L. (1987). *Invitations to science enquiry* (2nd ed.). Thornhill, Canada: S17 Science.

Miller, J. D. (1998). The measurement of civic scientific literacy. *Public Understanding of Science, 7,* 203–223.

Miller, J. D. (2010). Civic scientific literacy: The role of the media in the electronic era. In D. Kennedy & G. Overholser (Eds.), *Science and the media* (pp. 44–63). Cambridge, MA: American Academy of Arts and Sciences.

National Education Association of the United States. (1894). *Report of the committee of ten on secondary school studies.* Retrieved from https://archive.org/details/reportofcomtens00natirich

National Science Board. (2010). *Science and engineering indicators 2010.* URL (consulted 16th January, 2010) http://www.nsf.gov/statistics/seind10/

National Science Foundation. (2014). *Science and engineering indicators 2014.* Arlington, VA: National Centre for Science and Engineering Statistics.

NGSS Lead States. (2013). *Next generation science standards: For states, by states.* Retrieved from www.nextgenscience.org/overview-0

OECD. (2015). *PISA 2015 draft science framework.* Retrieved from http://www.oecd.org/pisa/pisaproducts/pisa2015draftframeworks.htm

Ogawa, M. (2013). Towards a 'design approach' to science communication. In J. K. Gilbert & S. M. Stocklmayer (Eds.), *Communication and engagement with science and technology: Issues and dilemmas* (pp. 3–18). New York, NY: Routledge.

Paisley, W. J. (1998). Scientific literacy and the competition for public attention and understanding. *Science Communication, 20,* 70–80.

Poliakoff, E., & Webb, T. L. (2007). What factors predict scientists' intentions to participate in public engagement of science activities? *Science Communication, 29,* 242–263.

Porter, G., & Friday, J. (1974). *Advice to lecturers: An anthology taken from the writings of Michael Faraday and Lawrence Bragg.* Sussex, UK: The Royal Institution.

Powell, M. C., & Colin, M. (2008). Meaningful citizen engagement in science and technology: What would it really take? *Science Communication, 30,* 126–136.

Power, M., & Dalgleish, T. (2008). *Cognition and emotion: From order to disorder* (2nd ed.). East Sussex, UK: Psychology.

Rennie, L., Goodrum, D., & Hackling, M. (2001). Science teaching and learning in Australian schools: Results of a national study. *Research in Science Education, 31,* 455–498.

Rennie, L. J., & Stocklmayer, S. M. (2003). The communication of science and technology: Past, present and future agendas. *International Journal of Science Education, 25*, 759–773.

Rennie, L. J., & Williams, G. F. (2002). Science centres and scientific literacy: Promoting a relationship with science. *Science Education, 86*, 706–726.

Rutherford, F. J. (1997, October). *Sputnik and science education.* Paper presented at the symposium reflecting on sputnik: linking the past, present and future of educational reform. Washington, DC.

Searle, S. (2014). *Developing an evidence base for science engagement.* Canberra, Australia: Australian Government.

Shen, B. S. P. (1975). Science literacy and the public understanding of science. In S. B. Day (Ed.), *Communication of scientific information* (pp. 44–52). Basel, Switzerland: Karger.

Silvia, P. J. (2006). *Exploring the psychology of interest.* New York, NY: Oxford University.

Stocklmayer, S., & Gilbert, J. K. (2002). New experiences and old knowledge: Towards a model for the personal awareness of science and technology. *International Journal of Science Education, 24*(8), 835–858.

Stocklmayer, S. M. (2013). Engagement with science: Models in science communication. In J. K. Gilbert & S. M. Stocklmayer (Eds.), *Communication and engagement with science and technology: Issues and dilemmas* (pp. 19–38). New York, NY: Routledge.

Stocklmayer, S. M., & Bryant, C. (2012). Science and the public—What should people know? *International Journal of Science Education: Science Communication and Engagement, 2*, 81–101.

Sturgis, P., & Allum, N. (2004). Science in society: Re-evaluating the deficit model of public attitudes. *Public Understanding of Science, 13*, 55–74.

Sturgis, P. J., & Allum, N. (2001). Gender differences in scientific knowledge and attitudes toward science: Reply to Hayes and Tariq. *Public Understanding of Science, 10*, 427–430.

The Guardian. (2014, May 6). *OECD and Pisa tests are damaging education worldwide – Academics.* Retrieved from http://www.theguardian.com/education/2014/may/06/oecd-pisa-tests-damaging-education-academics

The Royal Society of London. (1985). *The public understanding of science.* London, UK: The Royal Society.

Trench, B. (2006, May). *Science communication and citizen science: How dead is the deficit model?* Unpublished manuscript, based on a paper presented to scientific culture and global citizenship, ninth international conference on public communication of science and technology (PCST-9), Seoul, Korea.

Walker, G. (2012). *Motivational features of science shows.* Unpublished PhD thesis. Canberra, Australia: The Australian National University.

Wynne, B. (1991). Knowledges in context. *Science, Technology and Human Values, 16*, 111–121.

Wynne, B. (1992). Misunderstood misunderstanding: Social identities and public uptake of science. *Public Understanding of Science, 1*(3), 281–304.

Wynne, B. (2006). Public engagement as a means of restoring public trust in science—Hitting the notes, but missing the music? *Community Genetics, 9*, 211–220.

Ziman, J. (1991). Public understanding of science. *Science, Technology and Human Values, 16*, 99–105.

Reinvigorating Primary School Science Through School-Community Partnerships

Kathy Smith, Angela Fitzgerald, Suzanne Deefholts, Sue Jackson, Nicole Sadler, Alan Smith, and Simon Lindsay

Abstract Many primary school teachers, when supported by opportunities that assist them to reframe their thinking about the nature of science, appear to demonstrate a capacity to willingly use new perspectives to reconsider science learning and teaching. In particular the need for science to be explored as a human endeavour and the need to generate for students reason to seek understanding, to make sense of and communicate thinking about phenomena and experiences. To this end primary teachers value science learning situated within experiences that are personally meaningful and contextually relevant to their students, often producing opportunities to invite perspectives and achievements from sources outside the school to broaden science learning beyond the confines of the classroom. When established effectively such partnerships can potentially enable students to engage in and develop an understanding of science as a process of investigation and collaboration dependent upon the social construction of knowledge. Through an exploration of three case studies, we demonstrate situations where primary teachers and schools intentionally take steps to ensure their students have a sense of connectedness to their local community and environment by developing mutually beneficial learning relationships with both formal and informal science partners. By doing so these schools actively broaden the primary school science curriculum to include aspects of contemporary science with a particularly strong emphasis on social and emotional

K. Smith (✉) · A. Fitzgerald
Monash University, Melbourne, VIC, Australia
e-mail: kathleen.smith@monash.edu; angela.fitzgerlad@monash.edu

S. Deefholts · S. Jackson
St Joseph's Primary School, Crib Point, VIC, Australia
e-mail: sjackson@sjcribpoint.catholic.edu.au

N. Sadler
St Aloysius Primary School, Queenscliff, VIC, Australia

A. Smith
Holy Child Primary School, Dallas, VIC, Australia
e-mail: asmith@hcdallas.catholic.edu.au

S. Lindsay
Catholic Education Office Melbourne, Melbourne, VIC, Australia
e-mail: SLindsay@ceomelb.catholic.edu.au

© Springer International Publishing AG, part of Springer Nature 2018   87
D. Corrigan et al. (eds.), *Navigating the Changing Landscape of Formal and Informal Science Learning Opportunities*,
https://doi.org/10.1007/978-3-319-89761-5_6

aspects of learning. The result is a wider range of learning outcomes than were ever intended or anticipated for students, teachers and the community in general. Finally, the chapter identifies the characteristics that make school-community partnerships educationally valuable for science learning and teaching.

Keywords Science teaching · Science learning · School-community partnerships

Introduction

In a school-based meeting, a group of primary teachers shared their stories about an *inquiry unit*[1] they had recently completed with their Grade 1 and 2 classes (7–8 year olds). In this unit, history and science curricula were woven together in ways that linked directly to student wonderings and questions about 'change'. The classes explored change over time in the areas of technology, materials and the local environment by drawing on knowledge and experience from a range of sources, seeking out and working collaboratively with groups and organisations outside the school. Grandparents were invited as guest speakers and worked regularly in class with the students, the classes visited local historical sites and accessed museum education staff, who answered their questions as well as prompted further questions and ongoing inquiry about a range of issues. The enthusiasm of these teachers was infectious; they were thrilled to see opportunities to actively build connections between students and the local community and could see possibilities for links to wider global issues. They were finding ways to illustrate and engage students in science as a human endeavor—a way of thinking and acting where science knowledge and action is intrinsically linked with people and place. As one teacher commented, they were beginning to see a "bigger picture".

The above anecdote illustrates that these primary teachers found ways to move beyond their physical surroundings and link science learning with elements beyond the classroom. These teachers worked within Catholic Education Melbourne (CEM), an educational sector in Australia that has developed and articulated a clear vision for science education. This vision is designed to explicitly and actively support teachers to engage their students in science in both in school and out of school settings so that they experience and come to know science as a process of investigation and collaboration dependent upon the social construction of knowledge.

This chapter seeks to explore the conditions that enable primary teachers to reinvigorate school-based science education by creating mutually beneficial learning partnerships with local community groups and other organisations. The chapter initially examines the importance of science education that is well supported by strategic initiatives designed to empower teachers to reconsider science learning and teaching. Next, three cases are presented, each written by teachers as a way of

---

[1] 'Inquiry units' in Australian primary schools often refer to a multi-domain inquiry approach to curriculum planning designed to foster meaningful links across curriculum areas in order to enhance students' learning across subject areas. In these units students are encouraged to question their world, and use these questions to investigate a range of phenomena and how those phenomena impact on them.

capturing, portraying and sharing their professional knowledge about establishing school and community science partnerships. Each case illustrates situations where primary teachers and schools, with the backing of their educational sector, intentionally take steps to ensure that science teaching draws on mutually beneficial learning relationships with both formal and informal science partners to enable students to have a sense of connectedness to their local community and environment. The cases share experiences and insights in ways that identify the specific needs that became catalysts for establishing connections and the types of enablers that facilitated and ensured contextually rich learning. In all cases conditions were created that strengthened and further developed science education, producing a wider range of learning outcomes than was ever intended or anticipated for students, teachers and the community in general. When primary teachers, sector leaders and partnership organisations worked together in ways described in these cases, primary science education was reinvigorated.

The Role of Sector Support

In 2008 the CEM produced a vision for contemporary science education by articulating ten desired outcomes for all students. This grew from the work of a Science Reference Group which brought together teachers, principals, policy makers and academics to elicit a set of values around student learning in science so that the sector could explicitly identify what it "valued from a science education" (Lindsay, 2011, p. 7). Alongside this educational vision was the CEM's *Outward Facing Schools Charter*, an initiative again developed with teachers, which asserts that schools are well placed to open their doors to their communities to enhance learning for young people as well as strengthen connectedness and a sense of belonging. These sector statements were further supported by a range of strategic initiatives in science education that aimed to empower teachers to effectively identify student needs, contextualise opportunities for learning, and explore the potential for collaborative partnerships to enhance science education. In this context, science learning was expected to recognise and value local community and wider global initiatives as a way of providing contextually rich dilemmas for student learning and action.

> The CEM believes that learning in science is best when it connects strongly with communities and practice beyond the classroom. We commit through our underpinning values to building a culture of learning together in community, through collaboration, partnerships and life-giving relationships, which enable all to flourish together. We hold that links between the classroom and the local and broader community lead to students developing a rich view of the nature of science as it matters to them in the modern world. School community partnerships help to address ideas of relevance and connectedness within science learning for students, and this engagement with a more authentic version of science provides an entry point into deep learning of scientific concepts and practices. Learning within the community often provides students with opportunities to solve real issues and problems within society, and opens up engagement with social and ethical implications of their decision-making. In this way, school community partnerships in science offer learning that

has something at stake for the students, a set of consequences, which adds an edge to their learning. School community projects also provide potential contexts for innovation in science learning and teaching. The CEM believes that supporting groups of schools to trial, test and pilot novel programmes is vital to generating and sustaining innovation in science education across the whole system. Increased collaboration between Catholic primary and secondary schools and science organisations, universities, business and industry and other organisations open up possibilities for testing the boundaries of traditional learning. (S. Lindsay, Manager Improved Student Learning Outcomes CEM, personal communication, June 12th, 2015).

What this means is that science is modeled as the collaborative work of many people and this requires teachers to think differently about the nature of science and their role as science teachers. To achieve these intentions schools naturally look to the CEM for support and this required the development of a range of strategic initiatives to support schools in working towards achieving this type of science education.

One such strategic initiative was to provide opportunities for teacher professional learning. The earlier anecdotal story of teachers sharing their work was situated within a professional learning programme, *Contemporary Approaches to Primary Science* (CAPS). CAPS provides a rich example of a CEM initiative designed to support primary teachers to reframe science education and open up possibilities for school-based change. As an initiative that sits within and is highly valued as part of the overall sector vision for science education, the programme works to empower primary teachers to think and work differently with science (Smith & Lindsay, 2016) and investigate scientific literacy as a means of engaging students with contemporary science in the twenty-first century. All aspects of this professional learning experience positions teachers as key decision makers in terms of what matters in science learning and teaching. They work together in school teams and plan action research projects that reflect the often unique needs of their teaching context. Opportunities are provided for teachers to engage with a range of learning experiences that connect science skills and knowledge across curriculum areas. Most importantly, the programme actively models the place and value of collaborative science partnerships as a way of enhancing science learning and teaching.

The CEM has made significant and ongoing investment in building teacher capacity to develop collaborative relationships with external partners. Through a clearly stated vision for science education, strategic approaches to school support, and purposeful professional learning opportunities, primary teachers in particular are supported to recognise that science partnerships can enhance science learning and teaching as well as potentially enhance accountability and responsibility for capacity building and student achievement. To date the sector has achieved a number of valued outcomes as summed up below.

We are seeing the outcomes of these partnerships in terms of increased student engagement with learning, not just in science, but also across learning areas and attitudes towards schooling. We see particular gains in learning in traditionally marginalised students groups—those with behavioural and attention issues within normal classrooms. Engaging in learning in different environments using different skills and knowledge sees these students grow in confidence, self-concept and develop a sense of agency in their learning.

We are seeing impact at the whole school level as well. Schools have been able use the project grants as a stimulus to seek further funding opportunities external to the CEM. Schools have used project grants to develop local community projects which help to define the learning narrative of the school, and this assists the schools to bring the parents and community inside the school. (S. Lindsay, personal communication, June 12th, 2015)

The Case(s) for School-Community Partnerships

Such partnerships appear to be making some real headway in reinvigorating school-based science education. Yet what types of partnerships do schools choose to pursue and what holds these partnerships together so that invested effort is matched with quality learning? Science partnerships have been actively undertaken by a number of Catholic primary schools within the CEM sector and the following cases provide the professional insights of four primary teachers: two members of a school leadership team, an experienced primary science educator and a primary school principal, all of whom have been involved in the development of school-community partnerships to enhance science learning in their school. These teachers share their stories about professional knowledge of practice and convey the approaches and outcomes, both anticipated and unforeseen, that emerged from their work. The stories suggest that establishing productive partnerships is initially catalysed by certain events that create a recognised need for action and change. The ensuing partnerships are developed by some type of enabler that recognises and links contextually relevant opportunities to promote meaningful student learning. Strategic support becomes essential to initiating and maintaining the conditions for effective collaboration. For many of the teachers, these experiences have been transformative for themselves, the students and their school community as a whole.

Case 1: Making The Most of the Mangroves
Suzanne Deefholts and Sue Jackson

Context

St Joseph's Catholic Primary School is situated in a small coastal community located in South-Eastern Australia. Enrolments are drawn from six townships as well as from the families situated at the nearby Australian Defence Forces (Air Force, Army and Navy) base. Currently, we have an enrolment of 178 students with approximately 30% having one or more parent employed in the Defence Forces. Due to the transient nature of this population, and the occasional movement of local families, student retention rates across the 7 years of schooling (5–12 years) are at approximately 50%.

What did Science Education look like at Our school?

- Science was taught as a specialist subject once a week in gender groups.
- Science curriculum was not linked to classroom programming.

- Science was the sole responsibility of the Science Leader and no emphasis was placed on building the capacity of others to teach science.
- Each term focused on a different aspect of the state-based curriculum: biological science, physical science, chemical science, and earth and space science.
- Science lessons were activity based, but without an allocated physical space this did not allow for ongoing scientific observations or investigations.

In 2010, the school's leadership team was part of a professional learning initiative, the *Contemporary Learning Project,* facilitated by the CEM. As part of this project, the staff began to reflect on current learning and teaching practices and identified the importance of supporting teachers to further develop their pedagogical skills to enhance science learning.

A change in thinking

The following year, with the support of the school's leadership team, four teachers attended another CEM initiative, the *Contemporary Approaches to Primary Science* (CAPS) programme. To ensure the success of the programme, the school chose to send a wide representation of staff including the deputy principal, science leader, specialist PE teacher and an experienced teacher, all of whom were also classroom teachers. The CAPS programme was not a one-size-fits-all, but allowed for and encouraged teachers to develop their own ideas and understandings around science education. As part of the programme, teachers were asked to engage in an action research project. Through discussions as a team and with the programme's facilitators, the school decided to explore the possibility of providing a contemporary approach to learning science for our students that incorporated real-life scientific research and exploration. We wanted authentic learning opportunities based in our context that utilised the local environment.

A stand of mangroves lines the shore of our community and is within walking distance of the school. These mangroves represent some of the most southern extents of mangroves in the world and they are a keystone species in the local aquatic ecosystem. It was decided that these mangroves could provide a compelling environment to stimulate contemporary science learning for our students. While teachers quickly recognised this as a great pedagogical opportunity, equally they recognised that they did not necessarily have the expertise or in-depth knowledge required to ensure a rich learning experience.

Beginning the partnership journey

We initially formed a partnership with CEM and relied heavily on Simon Lindsay, the science team leader at the time, to assist us in creating partnerships with experts who had the knowledge we were lacking.

In the early stages, a number of students who lived on the Defence Force Base recognised that there was a mangrove habitat situated close to their homes. We contacted the base and arranged for students and teachers to visit, but it quickly became apparent that the Defence Force personnel who spoke to the children had limited knowledge of the ecological importance of the mangroves.

Simon then introduced the school to Dr. Tim Ealey, an internationally acclaimed scientist who has devoted his days to reviving a fragile bay in South-Eastern Australia by regenerating mangrove colonies. Dr. Ealey became instrumental in providing teachers and students with relevant background knowledge and scientific understandings. He identified a geographical area in which our students could contribute to long-term change and supported school staff to plan an excursion to a nearby coastal community. On this excursion, students had the opportunity to put their newly attained knowledge and skills into action by planting mangrove seedlings grown by Dr. Ealey, then collecting seeds to begin their own propagation cycle for replanting the following the year. Dr. Ealey, in turn, assisted in the formation of a partnership with a local association advocating for the protection of seagrass.

The support of experts external to the school inspired students to take action in highlighting the importance of the mangroves in our local environment to a local and global audience. Some of these actions included the creation of a blog, pamphlets delivered to the local community, information books for lower level students and the designing of a t-shirt, which was worn on the mangrove planting day. As teachers across the school developed confidence with this approach to science learning and teaching, other community partnerships began to form around the Mangrove Regeneration Project. In addition, our principal and two teachers travelled to Lombok, Indonesia in early 2012 for a 6-week language immersion course. While there, another contact from CEM assisted in linking our school with a sister school in Lombok who were also in close proximity to shoreline mangroves. With the assistance of a local Indonesian marine biologist, Hani Nusantari, students continue to correspond in English and Indonesian by sharing photos and stories of their local marine environments.

Building further partnerships by embedding successful practice

After 5 years, the Mangrove Regeneration Project continues to gain momentum, with community partnerships being built upon and added to. Currently, in addition to the community partnerships already mentioned, we are working with The Dolphin Research Institute, primary schools located across urban Melbourne as well as in the Far North of Australia, and Kids Teaching Kids, an organisation aiming to inspire future environmental leaders (http://www.kidsteachingkids.com.au/). The success of the programme can be attributed to the school's continued participation in the CAPS programme and the support of school leadership. Over this time, changes in teaching staff have brought new and innovative ways to use the mangroves and the coastal environments to enhance science learning and teaching. To ensure the sustainability of the programme, the school has chosen to make this project a focus for Year 3 and 4 students (8–10 year olds).

Through the success of the mangroves project, teachers had seen firsthand how the knowledge and experience of experts could enhance and deepen student learning. This gave teachers the confidence to explore other projects and partnerships, and with the support of leadership a student-led café was created. During the planning phase, partnerships were made with a variety of stakeholders, such as non-governmental organisations, local governmental health bodies, youth service providers, other schools, a landscaper and local hospitality-focused businesses. Students continue to learn in and through the cafe on a fortnightly basis, serving and making further connections within our own school community as well as with local groups such as local council, church groups, community service clubs (e.g., Probus, a worldwide club for retired or semi-retired business or professional people), gardening groups and an ongoing and growing relationship with CEM.

Sharing the journey through creating opportunities

Further opportunities and partnerships have developed since 2010. Staff shared their experiences as a model of professional learning for other CAPS groups, CEM and principal networks. Students have had a number of opportunities to share their knowledge and actions with local gardening groups, businesses, other schools and environmental networks.

When planning, teachers now seek opportunities for community partnerships as they recognise them as a valuable means of supporting the development of deeper learning and greater student engagement. Our community partnerships have ensured that quality learning has been connected to a local and global scientific context. This has allowed students to apply skills and knowledge to take action in a meaningful way. By taking on the role of the scientist through investigating, recording and reporting their findings, they are creating new knowledge to share in a global arena.

Case 2: Walking the Talk
Nicole Sadler

Personal philosophy

My philosophy of teaching and learning, indeed, all that I am as a person, is a product of all my experiences and the people who have shared these with me over my lifetime. From a personal perspective, my Catholic school upbringing served me well for making positive life choices rooted in social justice and for 'the common good'. From a professional perspective, the most influential of these experiences is the 9 years I spent as a teacher in a remote region in North Western Australia known as The Kimberley. It is this context that shook my foundations and forced me into realisations that continue to shape my philosophies and pedagogies. The Kimberley awakened in me a childlike awareness of place, community and human purpose. I say 'childlike' because I now know that children are essentially born with a clear and unencumbered sense of place, community and purpose. It is only through modern 'western' socialisation and, sadly, through westernised schooling, that these innate longings become foggy and dulled.

Defining experiences

Living in the Kimberley region in an Aboriginal (First Peoples or Indigenous Australian) community was my first experience of desert living. Each class had at least one Aboriginal teacher aide who would interpret language if necessary, teach traditional language and tell stories about local culture, and every week we would spend at least one day 'going bush' (spending time in the local environment) with our class. I learned more about the natural world in just one year from those children than I had ever learned in my own 'western style' schooling years. They didn't always turn up on time and they didn't achieve many national testing outcomes, but those kids knew their place, they knew their land and they sure knew their purpose in relationship with the land. Thankfully I was alert enough to recognise this and value it and I hope in my short time there I believed in them enough that they might hold on to it all in spite of systemic and government policies that have since conspired to erode these ancient wisdoms. After 5 years in the Kimberley, my mentor and friend, Ivy told me, "You sad for your place Nicole. Go back home to your place but come back soon". She was right and that is what I did.

After some time at Scienceworks, a science museum in South Eastern Australia, and a second stint in the Kimberley region, fate again intervened and I landed a job at 'my place', St Mary's Catholic Primary School in an urban area of South Eastern Australia. I had been born and I grew up in this area, I had even attended St Mary's as a child. I believed life had come full circle and this was a cosmic message that everything I'd learned in the Kimberly needed to be put into action. Fortunately I had a very supportive principal and a patient and open-minded co-teacher who both encouraged my ideas and schemes. The time was ripe to test my philosophies.

Context

The school environment at St Mary's had become a typical inner suburban concrete courtyard. The only greenery in the school grounds was a small garden that was overgrown with invasive plant species and sick-looking fruit trees. The school grounds did not offer the 'wonder filled' experiences I was looking for to ignite the students' imaginations. But the suburb is situated on a small peninsula that extends into a bay. The bay borders the town on three sides and offers a myriad of locations to explore and learn about the natural world. One of the locations I knew well from my childhood is an environmental area known as the Rifle Range Reserve.

Now mostly a housing estate abutting a state and local government managed wetlands (known as the Jawbone Wetlands), the Reserve was an active rifle range when I was growing up. No-one was allowed into the Reserve but the shooters, so the adjacent coastline had been kept in a pristine state for over 100 years. This location was just 4.5 km from the school and I knew it was an environmental sciences goldmine. How better to get there than to introduce a bicycle programme? So, with support from the Assistant Principal and some eager parents, we did.

Seeking and fostering partnerships

Mode of transport issues solved and parent helpers established, we scheduled regular visits to the Jawbone Wetlands to engage in environmental science education. At the same time, I knew that I would need some expert support in helping students to interpret the wetlands and apply some scientific thinking to their wonderings. So who better to call upon than the locals who know it best—rangers from two relevant local and state government organisations.

The rangers were not only happy to become involved in our fledgling programme but they offered to connect our school with other organisations that had interest and funding (at the time) to support such projects. Valuing the place and nurturing community relationships saw the school receive a number of community grants that allowed us to grow the environmental science education programme on site at school, especially for younger classes who could not ride.

One grant allowed us to set up a native plant greenhouse and nursery and another bought hi-tech digital microscopes, water testing and macro invertebrate study kits. Another grant enabled us to set up a family herb and vegetable garden. Later we were able to develop the vegetable garden to incorporate chickens. Yet another community grant saw us redesign the little-used school garden along Japanese reflective garden lines albeit using plants indigenous to the local area. In the short space of just 3 years, the programme grew from a few outdoor education visits to the wetlands to a full-blown, whole school environmental science education programme.

More than ten mutually beneficial community partnerships were established, which included local and state government groups, not-for-profit and non-governmental organisations, corporations, national initiatives and private enterprises. Most importantly, the school's parent community established an enduring Parent Environmental Group. Some partnerships were short lived as their goals and needs were met, others endure and no doubt even others will emerge.

Considering the value

Students' sense of their place, their community and their ability to engage meaningfully in both of these, appear to have been positively influenced by this programme. Students from St Mary's have been invited to speak at state and national conferences and seminars. Their workshops have been recorded so that other schools can examine and learn from these models. They have been invited to collaborate in scientific studies, for example, of the effects of the North Pacific Sea Star in the bay and they have been nominated twice for state-level environment awards for their work. We know from parent feedback that students from St Mary's who engaged in this programme have gone on to not only choose science pathways in secondary schools but to also do extremely well in these subjects. Furthermore, there is evidence that some students who do not excel in traditional pen and paper type testing, but who experience authentic success in their learning through the environmental science programme, go on to achieve higher scores and improved outcomes in the years following this engagement.

Having seen the effects of innovative, authentic learning on the attitudes, knowledge and skills of students, there is no way that I could possibly begin to plan for science learning and teaching by thinking otherwise. In my current position at St Aloysius Catholic Primary School located in a regional part of South Eastern Australia, there is widespread staff and principal support to take these ideas and RUN! This year has seen the implementation of studies of the local waterway known as Swan Bay and we have already invited a number of external participants to collaborate in sharing knowledge and skills with the students: parent professionals, state government bodies, the local naturalist club, and Indigenous elder Uncle Dave. I believe that St Aloysius, with a small student population and close-knit parent community, is an even richer ground for fostering sense of place, community and purpose and I look forward to learning about this new 'home' of mine.

Case 3: Enhancing Learning in Unexpected Ways
Alan Smith

Context

Holy Child Catholic Primary School, situated in an urban area of South Eastern Australia, is a rich and diverse community. Like many schools in working class suburbs, we have a community who has come from many parts of the world with a wide range of stories about their lives before settling in Australia. Over 90% of our families are from non-English speaking backgrounds and while many struggle with learning English, they are often fluent in one or more other languages. The children of our families are usually bilingual and often translate for their parents. Many families in our school have deep cultural heritage going back thousands of years. Importantly, those with a Middle Eastern background have often had limited access to formal education in their own countries due to lack of opportunity through war, geographical location or the need to prioritise contributing to the family income over schooling. On arrival in Australia, these families are confronted with a culture that places a high value on formal education and family engagement with schools. This is challenging for those inexperienced with such expectations.

Listening, looking, learning

To improve connections between the school and families, the notion of a HUB was conceived as a place of learning for our families, students and the wider community. The HUB was designed to provide a place where groups could meet and access opportunities for informal learning, such as social and discussion groups, as well as more formal learning opportunities (e.g., registered courses).

When I first started as the principal I worked with the Family School Partnership Convener (FSPC), a colleague whose role was to support schools to improve their community relationships and we asked parents what they felt could be done to build their capacity to engage in school activities and support their children's learning. We also met with each cultural group represented within the school (e.g., Vietnamese, English, Iraqi, East Timorese, etc.) and overwhelmingly the discussions indicated that parents wanted access to English classes. So this became our initial focus, followed by the provision of computer skills then employment pathways. With this information and vision we explored opportunities to work with local training organisations and local government, and both parties provided local residents with no cost courses in information and communication technologies and English.

Around the same time, I participated in a study tour in the United States with CEM to further explore family-school partnerships. We visited a range of schools that worked closely with local community groups to enhance student learning opportunities. This experience opened up conversations and opportunities for further collaboration with the CEM around partnerships. As a result, I was invited to contribute to the writing of the *Outward Facing Schools Charter*, a CEM document developed in consultation with a number of agencies, asserting the importance of school-community partnerships in contributing to successful learning experiences. This experience coincided with a large-scale Australian government funded building initiative, which supported our school to undertake major building works. This required two existing portable classrooms to be relocated and instead of disposing of them, they were relocated within the school grounds for use as spaces for community learning activities.

These classrooms were to become known as the HUB, a term connected with an initiative of an Australian-based philanthropic group, the Scanlon Foundation, which supports projects that lead to improved social cohesion. The Foundation had identified that the urban area in which the school is located had one of the lowest participation rates in Australia for the education of preschoolers and wished to make a financial investment to improve this situation. As a result the Foundation was seeking to financially support the development of 'Early Years Hubs'. Given CEM's priority to build effective links with community and the school's need to engage our parents, we recognised the concept of a HUB as one that could promote beneficial community links and parent involvement. The school's FSPC connected with local government and our school began to work with the Scanlon Foundation. The Foundation provided money for the employment of a HUB coordinator and has continued this commitment.

Making and maintaining partnerships

The first supportive partnership created through our HUB was with mothers from our community and the Year 4 children (9–10 year olds). A produce garden was planted in close proximity to the HUB to encourage students and parents to work together growing vegetables, herbs and fruit. The location also enabled these groups to make use of the HUB's kitchen facilities. In the early stages of this project, I showed some of the mothers the seedlings I had purchased and they laughed politely with the implication that I had no idea what to look for in terms of appropriate plants! These mothers, who were participating in the HUB English classes, offered their time and assistance to select, plant and help maintain the growth of suitable produce. The mothers worked in the garden every morning explaining to children, in slowly improving English, how to plant and nurture seeds.

Since 2013, the children and parents have supplied seasonal vegetables for cooking classes held in the HUB. A community kitchen programme has been established with the support of local council to promote healthy eating practices to our Year 5 (10–11 year old) students. This support enables our students to benefit from the informal agricultural and botanical knowledge these parents have developed from their personal experiences and traditions of working closely with the land. The knowledge the parents share enables the students to make connections between the processes involved in growing, cooking and eating produce from our gardens. The children develop understandings around ideas such as relationships and survival by observing seasonal changes and the impact of these changes on conditions for plant growth, including appropriate soil temperature and moisture, exploring companion planting, and experimenting with physical and chemical change as they prepare meals in the cooking programmes. The parents gain the opportunities to talk and improve their spoken English, sometimes being challenged to explain difficult concepts in simple language for the children to understand.

Another significant partnership is with the Faculty of Education at the Australian Catholic University (ACU). This partnership has allowed our children and ACU pre-service teachers to form connections through science learning and teaching. Numerous benefits have resulted from our ACU partnership, for example, our children and their parents are exposed to people studying at university. They can ask about university life, further studies, and pathways to learning and post-secondary education.

Conditions for School-Community Partnerships

At a glance, these cases seem to tell three very different stories about the ways in which primary schools and classroom teachers seek to use school-community partnerships to bridge the gap between in school and out of school science learning experiences. On the surface it might seem like the natural environment is the common thread pulling these partnerships together but closer examination reveals that it runs much deeper than this. These three cases illustrate that while there is no single way to foster a successful school-community partnership, there are a number of components that can foster the right conditions (and is similar to the 3 "C"s coordination, customisation and connection identified in chapter "Viewing Science Learning Through an Ecosystem Lens: A Story in Two Parts" by Falk and Dierking). From these cases, the following four conditions emerged:

1. recognising the *need* for change,
2. someone or something *enabling* change,
3. seeking out the right *partner(s),* and
4. application of the learning from the partnership to promote further *growth.*

Each condition is explored in more depth below.

The Need for Change

These three cases each recognise that change was needed in the school or classroom, in one form or another, to better engage students and their families in the learning process. It is interesting to note that regardless of what this change was—a focus on contemporary approaches to science education (Case 1), forging stronger connections with place (Case 2), or supporting parents to join the school community (Case 3)—it was done in a way that was relevant and meaningful to that particular community. For example, the partnerships formed in Cases 1 and 3 sought to better engage with a diverse student population, while Case 2 identified the author's own values as a driver for a collaborative approach to practice. For all cases, there was recognition of wanting to engage students and their communities in more authentic ways of learning science. Equally, across the three cases, it was recognised that partnerships with individuals and/or organisations would be needed in the form of expert input, whether this be connected to the sharing of expertise, funds or both, to bring about more significant and long-lasting change. When context is utilised as an enabler of change, opportunities emerge that enhance science learning and teaching. Yet recognising the many unique opportunities context provides for learning can be challenging and so providing strategic support to enable alternative action is essential to ensuring that partnerships add value to student learning.

Enabling Change

In each of these cases, something enabled action that led to the desired change. In these instances, the initial enabler was the author of the case (Case 2), a professional learning programme (Case 1), and a working partnership between two education professionals undertaking complementary roles (Case 3). Over time this enabler played a key ongoing role, but other individuals and organisations moved in and out at different times to support and influence the change that was taking place. Beyond the human enablers, each case was also impacted by a set of enabling factors that were contextual, like the environment (e.g., access to mangroves, wetlands, vegetable gardens) or resources (e.g., bicycles, portable classrooms). These contextually relevant opportunities played a critical role in not only connecting the school with the community, but were utilised to provide meaningful learning and interactions for all involved.

The stories suggest that the catalysts for establishing productive partnerships range from individual values to targeted professional learning experiences, and create opportunities for teachers to think about and value the need for meaningful learning and capacity building to initiate change. Encouraging schools to become active players in establishing school-community partnerships is essential to the success of any such collaboration. Obviously teachers need to see a reason to invest their time and energies beyond an already busy and demanding teaching schedule. Finding the right partner is a critical part of the process.

The Right Partners

All three cases highlight that partnership selection really matters. It is evident that pursuing mutually beneficial partnerships with outside groups is an effective way of attending to the needs of both schools and communities, but the partner individual or organisation plays a critical role. Not only are they active participants in a learning experience that is often in its developmental stages, but they are also instrumental in knowledge sharing, assisting with identification of possible funding avenues and opening up opportunities for different partnerships. Interestingly each case highlighted the involvement of multiple partners from a variety of sectors and backgrounds. There seemed to be a ripple effect connected with the partnerships noted in these cases, with one partner leading to another to another in local (e.g., Case 1), national (e.g., Case 3) and sometimes even international settings (e.g., Cases 1 and 3). This is also illustrative of the reality that partnerships will differ in terms of the timeframe, objectives and strengths they bring, which makes this type of work dynamic and responsive to needs. Not all partnerships formed will, or need to, be long-lasting and transformative as each has different purposes to fulfill. Emerging from these stories is that strategic support becomes essential to initiating and maintaining the appropriate conditions for effective collaboration.

Further Growth

It is often hard to predict when a partnership is formed exactly what the outcome will be. There is a need for both parties to be open-minded about the possibilities. In each of these three cases, after the initial need was met, unexpected positive outcomes started to arise for all stakeholders. For Case 1, it was the chance to apply the learning that had taken place to a new location and to contribute to increased public involvement in the sustainability of the local area. For Case 2, it was an opportunity to experience partnership building with different members and organisations in an entirely new context and contribute to the work of these organisations as they strive to enact their agendas of achieving wider public engagement. And for Case 3, it was the emergence of a number of different dimensions that resulted from the formation of one key partnership. This partnership went beyond the initial intentions of school- and community-based growth to contribute to the enhancement of learning opportunities for future teachers.

All three cases revealed aspects of personal growth, and the teachers as well as students became involved in sharing the outcomes of the various partnerships in a variety of forums, such as through blogs and at conferences. These opportunities enabled the showcasing of models for and outcomes of such partnership formation for other teachers and students as a way of highlighting what is possible. The experiences that ensued from these school-community partnerships were transformative for many within the schools as well as the school communities as a whole.

For all involved the partnership needs to be lived and experienced, requiring a degree of flexibility to respond to concerns and issues as they arise. Investing in such a process essentially involves taking risks and dealing with the uncertainty of outcomes and a willingness to embrace both the intended and the unanticipated learning that emerges. In terms of thinking about sectorial support arising from these stories, it is anticipated that the value of these school-community partnerships within future science policy will move to supporting *school-to-school* projects in collaboration with community. The benefit of clustering and sharing knowledge and expertise between schools holds significant potential. There is also much potential for the sharing of community resources across schools within the community. This notion would consider the community as a hub, which can provide a range of shared services, such as health, education, library, business, and social support for schools. In this way, with this support from community, schools are able to provide opportunities for students that actually matter to them—opportunities that are grounded in their environment, their community, their world.

Conclusion

School-community partnerships can take many guises, but essentially the development and fostering of a relationship must start somewhere. The cases shared in this chapter illustrate how teachers identified the uniqueness of learning within and about place and created opportunities in their local context and beyond for rich science learning. Schools worked with the backing and support of their sector to collaborate with outside organisations and provide a range of learning experiences that promoted rich and contextually relevant learning. One of the most transformative aspects of this support was the degree of trust that CEM placed in school leadership and teachers to make decisions about the learning that mattered for their students. This trust enabled teachers to determine how they could best utilise the local environment and mobilise partnership opportunities to enhance student learning. Investing trust in teachers as professionals appeared to be a key condition that enabled these primary school teachers to find their voice as science educators, and become active decision makers about quality science teaching and the importance of contextually relevant learning. Primary science education was reinvigorated with connected and interactive partnerships that realised not only a purposeful vision for meaningful science learning and teaching but also enabled schools to become active participants in their local communities and position themselves as outward facing schools.

While these stories may illustrate ideal notions about school-community partnerships, it is important to note that they were not without their challenges: teachers needed to think differently about the role of context in science learning and teaching, support was needed to enable meaningful and sustainable change, and finding and establishing the right partnership was critical. Yet these challenges were overcome when a sector explicitly valued and strategically supported schools and teachers to play an important role in developing contextually relevant science education for their students. In these conditions partnerships were established that enabled students to develop rich understandings of the nature and place of science in their world. Students, teachers, parents and outside organisations worked together to co-create collaborations that provided opportunities for all parties to engage in meaningful dialogue and inform focused, considered, ongoing and sustainable action. Partnerships established in these ways benefitted all stakeholders, with schools undertaking clearly defined roles in the action of their community. Once this culture of collaboration is established, it becomes difficult to consider science learning and teaching in any other way—it is the school and the community sharing responsibility for quality science education, which can only lead to a positive future.

References

Lindsay, S. (2011). Scientific literacy: A symbol for change. In J. Loughran, K. Smith, & A. Berry (Eds.), *Scientific literacy under the microscope: A whole school approach to science teaching and learning* (pp. 3–16). Rotterdam, The Netherlands: Sense.

Smith, K., & Lindsay, S. (2016). Building future directions for teacher learning in science education. *Research in Science Education, 46*, 243–261.

Natural Disasters as Unique Socioscientific Events: Curricular Responses to the New Zealand Earthquakes

Léonie Rennie, John Wallace, and Grady Venville

Abstract This chapter examines earthquakes as a real-world, socioscientific issue to explore how schools, school curricula, school systems, and communities respond to the learning opportunities created by a natural disaster in the local or global community. We identified some of the issues that determine how different countries deal with earthquake preparation, response, and the factors that affect recovery. We then reviewed school-based, curriculum, and community responses to the Canterbury earthquakes in New Zealand as a case study. In the immediate aftermath, attention focused on the emotional support of students, teachers, and families, and efforts were made in combination with the community to return to normal schooling and curriculum stability in students' lives. We suggest that recovery from such natural disasters must be both flexible and integrated across curriculum and the community, drawing widely on available resources.

Keywords Science curriculum · Multi-disciplinary science learning ·
Socioscientific issues

In the Volcanoes and Earthquakes Gallery at the Natural History Museum in London a very large television continuously screens a montage of contemporary video visuals about the March 2011 Tōhoku earthquake in Japan and the following tsunami. Each of five separate screen segments simultaneously presents nearly 6 min of assembled footage (courtesy of the Tōhoku Broadcasting Company/TBC/JNN)

L. Rennie (✉)
Curtin University, Bentley, WA, Australia
e-mail: L.Rennie@curtin.edu.au

J. Wallace
OISE, Toronto, ON, Canada
e-mail: j.wallace@utoronto.ca

G. Venville
University of Western Australia, Crawley, WA, Australia
e-mail: grady.venville@uwa.edu.au

© Springer International Publishing AG, part of Springer Nature 2018   105
D. Corrigan et al. (eds.), *Navigating the Changing Landscape of Formal and Informal Science Learning Opportunities*,
https://doi.org/10.1007/978-3-319-89761-5_7

from different locations showing the actual quake and its immediate aftermath; a total of about 30 min of awe-inspiring, massive destruction. Half an hour is a long time for a museum visitor to sit and watch any exhibit, but the seating opposite the screen is well-used by visitors engrossed in this powerful imagery.

None of this visual montage is new; it has been seen many times since March 2011, but it is chillingly, breathtakingly real. Although the magnitude 9.0 Tōhoku earthquake was neither the largest this century nor the most deadly (the magnitude 9.1 Sumatra–Banda Aceh earthquake and tsunami in 2004 took the lives of around 230,000 people), it received extensive media coverage. Indeed, it was the first terrestrial event to be broadcast simultaneously on public and commercial television, radio, and live via the internet (Murakami, 2014). Information was accessible to people even without electric power through mobile electronic devices. Anyone with a digital communication device, anywhere in the world, could be connected to the ongoing disaster in real time.

This degree of connectedness reflects a new global reality. Once relatively local issues, like earthquakes, floods, hurricanes, and wildfires that concerned other people elsewhere, now touch us wherever we live. Our children are growing up in an increasingly technological environment, and as worldly citizens, will need the knowledge and skills to respond to both local and global matters. They will need more than just knowledge of the science that underpins natural disasters, they will also need an understanding of what can be done to minimise the risk of damage and injury, and to mitigate the ensuing circumstances. Further, the decisions made about preparedness, coping, and recovery do not rest on science alone, they involve emotional, social, economic, and political considerations. In the face of disasters like major earthquakes, such decision-making is sure to be contested and the outcomes will be felt over many years.

Here we join the ongoing conversation about how best to equip students to explore and understand socioscientific issues and events, and to experience the kind of connectedness that reflects life outside of school. In doing so we recognise that school science curricula, particularly in high schools, tend to focus narrowly on science concepts rather than encourage a broad-based focus on what is happening in the world outside. We also know that school systems do not adapt quickly to changing educational needs for a range of reasons, for example, the limits of time, structure, and school priorities that have more to do with bureaucracy than the higher level skills required by students in the twenty-first century (Schwarz & Stolow, 2006). Further, the dominant approach to high school science education is uni- rather than multi-disciplinary. Challenging the curriculum status quo—what Tylack and Tobin (1994) called the "grammar of schooling"—is not easy. Some of the answers, we believe, may be found by looking to the informal sector, which is "less wedded to traditional texts and much more engaged in context-based science … [it] can and does provide for disciplinary integration and a more holistic picture of what science is really like in the world outside of school" (Stocklmayer, Rennie, & Gilbert, 2010, p. 28). But we also know that an integrated or multi-disciplinary approach to the school science curriculum increases students' opportunities to engage in contextual, issues-based learning (Rennie, Venville, & Wallace, 2012).

In this chapter we take a closer look at how these three issues—the need for a more connected and responsive science curriculum, the relative inertia of the school sector compared with the informal sector, and the potential of integrated curricula—interact and collide in the face of a major natural disaster. We are interested in what happens in a school curriculum when a significant socioscientific event arising in the community is something that the school cannot ignore. What happens to curriculum in the face of a natural disaster like an earthquake, for example? How do schools equip students to live with and prepare for the "uncertain" possibility/probability of an occurrence or recurrence of such an event?

We chose to focus on natural disasters—earthquakes, tsunamis, hurricanes, wildfires, and other major life-threatening emergencies—as special cases of socioscientific issues. Such events are particularly important given that a significant portion of the world's population has recently experienced such events or resides within zones that are prone to be affected. These occurrences often have their share of controversy because they have deep social, political, economic, and emotional consequences with effects at the personal, local, national, and global level. Unlike many other socioscientific issues, natural disasters also have an important time dimension, both immediacy and recovery, and a high level of scientific unpredictability and uncertainty.

In the remainder of this chapter we return to the example of earthquakes as a real-world socioscientific issue with significant after effects to examine how schools, school curricula, and school systems respond to the learning opportunities created by a natural disaster in the local or global community. Why earthquakes? Every earth science curriculum, particularly at the senior level, has a section on earth movements, and every geography curriculum makes reference to landforms, so earthquakes are relevant to school curricula in at least two subject disciplines. While we appreciate that not every student or teacher will personally experience an earthquake or its aftermath, major earthquakes are significant events; they grab headlines world-wide, and a plethora of information about them is available from sources in the informal sector. Further, the kinds of school and community responses that are possible have parallels in other kinds of disasters that students may experience, such as wildfires, flooding, and extreme weather events (Mutch, 2014).

To give context to the following discussion, we begin by overviewing several recent international earthquakes to identify some of the issues that determine how different countries deal with earthquake preparation, response, and the factors that affect recovery. To address the curriculum focus of this chapter, we then turn to the effects of recent earthquakes in New Zealand and review school-based, curriculum responses to those earthquakes.

Learning About Earthquakes

In this section, we provide background information about major earthquakes in Japan and Nepal, both situated in zones of frequent seismic activity. This information also illustrates the contrasts between earthquake preparation and response in highly developed and underdeveloped countries. Next, we briefly review the longer term earthquake recovery process for earthquakes in Italy, Indonesia and New Zealand. In the subsequent section, we consider the case of the Canterbury earthquakes in New Zealand as a specific example of educational preparation and response to a major disaster.

The March 2011 Tōhoku Earthquake in Japan

The Tōhoku earthquake, also known as the Great East Japan Earthquake, occurred in the early afternoon of March 11, 2011. This magnitude 9.0 quake was caused by the movement of the Pacific tectonic plate pushing under the Eurasian plate. It was followed by a devastating tsunami that, among other destruction, severely damaged the Fukushima Daiichi Nuclear Power Plant resulting in a partial meltdown and extensive leakage of radioactive materials. About 16,000 people perished, mostly through drowning. Four years later, over 2000 people were still missing and nearly a quarter of a million people remained in temporary housing.

Japan has strongly enforced building codes, so buildings are expected to withstand earthquakes and they generally performed well. Regular exercises are held in schools and workplaces so that people know the best ways to protect themselves when an earthquake or tsunami occurs. The Japan Meteorological Association (JMA) has an earthquake early warning (EEW) system made possible because the preliminary p-waves (push or primary waves) from an earthquake travel nearly twice as fast as the more damaging, ground-shaking s-waves (shear or secondary waves), so the p-waves are detected sooner, with the time difference depending on distance from the epicentre and the intervening geology. Detection of the p-waves triggers a computer-based procedure that involves prediction of the magnitude of the earthquake and whether it is strong enough to initiate a warning (Yamasaki, 2012). JMA uses a geographically-focused, mass text messaging system to all cellular broadcasting-enabled mobile devices as the main warning method, supplemented by outdoor speakers, television and radio networks. Vervaeck and Daniell (2011) explained how the EEW enables trains, elevators, and factory processes to be slowed, or even shut down if there is time. People receiving text messages can seek safety. Yamasaki reported that during the 2011 Tōhoku earthquake, people in Sendai (129 km from the epicentre) received a warning 15 s before the arrival of s-waves, sufficient time to take cover, and in Tokyo (373 km from the epicentre), 65 s of warning time, sufficient time to shut down and evacuate 40 of the 42 elevators in the Tokyo's Metropolitan Government buildings, for example. These warnings are estimated to have saved thousands of lives.

The Tōhoku earthquake also generated a massive tsunami. Although the JMA issued tsunami warnings within 3 min of the earthquake, its severity caused data saturation of many sensors, resulting in initial under-reporting of its intensity (Japan Meteorological Association (JMA), 2013). This resulted in underestimation of the height of the tsunami waves, and thus possibly greater loss of life due to delay in evacuation, or failure to reach high enough ground. Further warnings containing heights re-estimated with additional data from ocean buoys were issued from about 25 min later. The scientists learned from these problems and by March 2013 a new warning system was put in place. Using principles from science, mathematics, technology, and engineering, the JMA enhanced its observation facilities and upgraded its algorithmic procedures for tsunami warning systems to avoid underestimation in the future. The JMA also prepared educational videos and distributed them to schools to educate the population to understand the improved warning system and assist disaster mitigation (Japan Meteorological Association (JMA), 2013).

The April 2015 Earthquake in Nepal

An earthquake of magnitude 7.8 occurred in Nepal around noon on April 25, 2015. It was caused by the Indian tectonic plate under-thrusting the Eurasian plate. The damage was compounded by aftershocks, including one of 7.3 on May 12. In the city of Kathmandu around 80% of buildings were damaged or destroyed and avalanches and landslides buried many villages in the Kathmandu valley. Kathmandu was a tourist destination not only for trekkers and climbers of Mt. Everest, but for its centuries-old temples, many of whose old brick structures crumbled into piles of dust. According to a report from the United Nations (UN News Centre, 2015b, June 9), around 8500 people perished and nearly 3 million were displaced.

Nepal is one of Asia's poorest countries with about half of its population living below the poverty line. Many buildings are very old and not built to withstand earthquakes. Overcrowding in cities and poverty means there are insufficient resources to construct earthquake-proof buildings, although with the assistance of the World Health Organization (WHO), some hospitals had been retrofitted and remained operational after the earthquake as they sustained much less damage (World Health Organization (WHO), 2015). Building codes, only recently introduced, are not well enforced (Poudel, 2015). In the Kathmandu Valley, a world heritage site, more than half of the unique temples, stupas and historic houses either collapsed or were seriously damaged (UN News Centre, 2015a, June 5). For the local Nepalese, whose daily worship involved these temples and other religious artefacts in the city and villages, this was a cultural disaster, compounding the emotional and social crisis of losing family and friends, homes, and often their livelihood. Poudel (2015) pointed out that international aid was essential to assist the Nepalese in the immediate aftermath of the earthquake, and it will continue to be essential to assist in the rebuilding process.

Variables Affecting Long-Term Recovery from Earthquakes

Earthquakes and their aftermaths, such as tsunamis and landslides, result not only in death, injury, and severe emotional stress, but damage to infrastructure, the environment, and industrial and agricultural activities. Not only may people lose family members, many more are made homeless and face loss of employment and livelihood. There is subsequent threat from shortages of food and clean water. These are social, cultural, environmental, economic, and political calamities and long-term efforts are required to address them. Frequently it takes years to recover some sense of normalcy, and there remain considerable psychological effects. Simplistically, we might expect the success of recovery to be a function of government-led administrative decision-making, but there are many variables involved. Some of these variables and concomitant complications were addressed by Alexander (2012) in his analysis of the short to medium term aftermaths of three medium-power earthquakes, those in L'Aquila, Italy (April 6, 2009, magnitude 6.3); Padang, Indonesia (September 30, 2009, magnitude 7.9); and the combined Canterbury earthquakes in New Zealand (September 4, 2010 and February 22, 2011 of magnitudes 7.1 and 6.3, respectively).

As Alexander (2012) pointed out, these earthquakes occurred in countries with very different geographical, economic, social, cultural, and administrative settings, and his examination of these variables illustrated very different approaches taken to recovery. Alexander explored the processes of decision-making in terms of international aid in disaster response, means taken to provide shelter for those left homeless, the clearing of debris, rebuilding, and efforts to re-instate the district's economy and employment. He looked particularly at the social aspects of recovery measures taken and contrasts between local interests and the consequences of broader government decisions. Alexander concluded: "In each case, although in different ways, the conclusion is that any explanation of how things proceed after disaster is difficult unless it takes full account of political realities" (p. 10/14). Emphasising that recovery is very long term, he also noted that "decisions taken soon after the disaster need to be examined in the light of their probable repercussions after the passage of decades" (p. 10/14).

Alexander's (2012) words reflect that recovery is a staged process. Mutch and Marlowe (2013) pointed out that all major disasters involve an immediate emergency response phase devoted to dealing with injuries and the provision of shelter, food and water, followed by a restoration phase, in which "people put their lives back together" (p. 390), and then a reconstruction phase. The length of these phases depends on the severity of the disaster, the resources available within communities, and, in many cases, assistance from international aid agencies. Obviously there are implications for schooling and curriculum. In countries where natural disasters are likely to occur, it might be expected that the curriculum will include education not only about the natural scientific phenomenon associated with such disasters, but also about disaster awareness, preparation, and risk reduction. Moreover, after the natural disaster event, a certain degree of flexibility and responsiveness might be expected in the curriculum to enable students to reflect, understand, and adapt to their new, post-disaster environment. In the following section we consider the specific case of the 2010–2011 Canterbury earthquakes in New Zealand, the nature and relevance of the pre-event curriculum, as well as the aptness and adaptability of the curriculum and educational processes in the context of the post-event community and environment. Further, we suggest that there are some aspects of the curriculum response in this example that may be transferable to other contexts.

Curriculum Implications of Earthquakes

Choice of the Canterbury Earthquakes in New Zealand

We chose the Canterbury earthquakes as our case study of curriculum response to a natural disaster for two reasons. The first was accessibility; we could find and read some education-based literature relating to the aftermath. The second reason was entirely educational: New Zealand has a national curriculum that is supportive of both integration and learning experiences outside of school, and also has a school-friendly approach to disaster awareness education. The following overview shows how a community-linked, integrated response to a significant disaster, such as an earthquake, is possible in such an environment.

The overall Vision of the New Zealand Curriculum (NZ Ministry of Education, 2007) is "Young people who will be confident, connected, actively involved, and lifelong learners" (p. 8). Each of these terms is explained, with "connected" defined as being: able to relate well to others, effective users of communication tools, connected to the land and environment, members of communities and international citizens. In addition, one of the Principles underpinning the curriculum is community engagement, requiring that "The curriculum has meaning for students, connects with their wider lives, and engages the support of their families, whānau [extended families], and communities" (p. 9). Clearly, community links are strongly encouraged. This is reflected, too, in the five Key Competencies: thinking; using language, symbols and texts; managing self; relating to others; and participating and contributing. It is difficult to imagine the development of the key competencies in other than cross-curricular contexts. The New Zealand Curriculum also explicitly encourages linking between learning areas, and descriptions of four of the eight learning areas suggest this potential:

> In **science**, students explore how both the natural physical world and science itself work so that they can participate as critical, informed, and responsible citizens in a society in which science plays a significant role.
>
> In the **social sciences**, students explore how societies work and how they themselves can participate and take action as critical, informed, and responsible citizens.
>
> In **mathematics and statistics**, students explore relationships in quantities, space, and data and learn to express these relationships in ways that help them to make sense of the world around them.

In **technology**, students learn to be innovative developers of products and systems and discerning consumers who will make a difference in the world. (p. 17)

Even though these descriptions are written as learner outcomes, rather than in terms of how teachers might assist learners to achieve them, these straightforward descriptions support the implementation of integrated topics across disciplines and making links to the world outside of school. Further, the Curriculum advocates Learning Experiences Outside the Classroom (LEOTC), and such learning is supported by its own website (eotc.tki.org.nz/LEOTC-home) and comprehensive guidelines (NZ Ministry of Education, 2009). For a topic like earthquakes, we might expect that science and social studies have great potential for integrating curriculum and making community links. Mathematics and technology are easily included by looking at ways of detecting and predicting earthquakes, and measuring earthquake magnitude, for example.

Much of New Zealand lies on the boundary between the Australian and Pacific tectonic plates, so New Zealand has a history of earthquakes and volcanism. New Zealand's civil defence authorities have long had programmes to assist the population to prepare for and deal with natural hazards, including earthquakes, volcanic eruptions, tsunamis, and severe weather events. In 2006, the New Zealand's Ministry of Civil Defence & Emergency Management (MCDEM) introduced a teacher resource with a primary school educational focus entitled "What's the Plan Stan?" using a dog called Stan as its icon (www.whatstheplanstan.govt.nz). The Teachers' Guide includes unit plans at the junior, middle, and senior primary levels with detailed learning intentions to fit the Health and Physical Education, Social Studies, Science, and English Learning Areas of the curriculum, and also a section on cross-curricular lessons for Technology and ICT, Food Technology, The Arts, Mathematics, and Learning Languages. The Guide also points out that disaster awareness education can be a context for an integrated learning approach and is suitable for use in LEOTC.

This brief outline shows that an integrated curriculum approach involving health, science, and social studies, including knowledge about natural disaster preparedness at school and in the community, is consistent with these ideas. However, any response must also take account of the emotional and contextual factors relating to the occurrence of a major disaster. In discussing school and curriculum responses to the Canterbury earthquakes, we are mindful of the significant emotional effects of these earthquakes on people, including teachers and students, in the local and surrounding areas. In broader geographic regions, however, earthquakes are noticeable and compelling natural disasters—they offer opportunities for integration across several curriculum areas in the context of community resources and this is the aspect on which we will focus. In the next sections we overview the Canterbury earthquakes then give attention to the educational responses to them. In doing so we recognise that the published reports present only a partial view of what happened in schools in the aftermath of the earthquakes.

The 2010–2011 Canterbury Earthquakes in New Zealand

The Canterbury earthquakes first struck on 4 September 2010 with an epicentre near the town of Darfield about 40 km from the city of Christchurch. The earthquake occurred on a previously unmapped fault line along the Australian-Pacific plate boundary. It had a magnitude of 7.1, occurred early in the morning and had many after-shocks. With few people about, there were no directly related deaths but there was considerable damage to buildings, including in Christchurch. Less than 6 months later, on 22 February 2011, came an earthquake much closer to Christchurch, of magnitude 6.3 and again followed by many aftershocks, some severe. This quake struck around lunchtime, with an epicentre very near the city centre, resulting in 185 deaths due mainly to collapsing buildings. Together with the numerous aftershocks (that continue for some time), the Canterbury earthquakes destroyed or seriously damaged many significant buildings in Christchurch. In surrounding areas, buildings and infrastructure such as roads, sewerage, and water supplies were further damaged or destroyed by liquefaction, a process in which muddy sediment welled up through ground fissures opened by the quakes and aftershocks, causing flooding and subsidence.

Educational Response to the Canterbury Earthquakes

The Canterbury earthquakes caused significant educational upheaval. New Zealand's Education Review Office (ERO, 2013) reported that all schools and early childhood services closed immediately after each of the two main quakes. In total, 215 schools were affected, but no child, student, or teacher in an education institution was killed or seriously injured. Within 12 days of the September quake, 99% of early childhood services and 98% of schools had reopened. Three weeks after the February quake, 62% of early childhood centres and 84% of schools were back in operation. Nearly 12,000 of 150,000 students relocated to other schools around New Zealand (the majority later returned), some schools shared buildings with others on a shift basis, and some students worked at home using learning hubs set up around the city, until they could return to school.

ERO (2013) collected and published "Stories of resilience and innovation in schools and early childhood centres: Canterbury earthquakes: 2010–2012" to report examples of innovative practices used by school personnel and develop recommendations to share the learning that had taken place. Four key themes guiding school activities emerged from the ERO synthesis: keeping children safe, supporting children's learning, supporting staff and families, and managing ongoing anxiety. In terms of learning and curriculum, ERO described the prevalent response in these terms:

> Teachers found that getting children and young people back into learning helped to normalise the situation for children and their families. The school's and service's curriculum

needed to be adapted to respond to the emotional and learning needs of their children and young people. (p. 1)

Overall, ERO concluded that:

The school was seen as a vital hub in the local community for not only the families attending the school, but also the wider community. Giving to others and connecting with the community was a very positive outcome of the crisis created by the Canterbury earthquakes. (p. 2)

The role of schools as important community hubs was also highlighted by Mutch (2014) in her synthesis of disaster events in Australia, Japan, and New Zealand. In times of crisis, schools are frequently places of physical shelter and relief. They are also "sites and facilitators of [disaster] preparedness learning and activities", the "first responders or post-event response centres", and in the recovery phase, "pastoral care and agency hubs for staff, students and families" (p. 19). Notwithstanding the important role of schools in all phases of the disaster process, Mutch found that there was little effort, internationally, to prepare them for this role. She noted that following the earthquakes in 2010 and 2011, the New Zealand Ministry of Education encouraged schools "to consider integrating disaster preparedness into the curriculum through health, science and social studies" (p. 15).

As noted earlier, New Zealand's civil defence authorities have programmes to assist its population to prepare for and deal with natural disasters, but school students' knowledge of these is variable. Finnis, Standring, Johnston, and Ronan (2004) interviewed 10- to 12-year-old children in a Christchurch primary school about their understanding of natural hazards. Children were aware of potential hazards, particularly storms and earthquakes, and safety behaviours were generally well understood. However, they reported rather poor preparedness plans in their households. Only about 20% knew about the Alpine Fault (associated with the Australian-Pacific plate boundary), and of those who did, only a third were aware that Christchurch could be affected by an earthquake along that fault. The authors recommended: "Awareness of the Alpine Fault and the impact of an event greatly needs to be increased considering the level of threat posed to Christchurch and the 'overdue' nature of an earthquake generated along the central Alpine Fault" (p. 19).

In March 2012, the MCDEM conducted a New Zealand-wide evaluation of "What's the Plan Stan?" across 1020 primary schools, finding that about three-quarters of schools were aware of "What's the Plan Stan?" but only 31% had made use of it (MCDEM, 2012, p. 12). Although it had been used in all curriculum areas, respondents to the survey indicated it was most often used in health and physical education (27%) and social studies (25%), followed by LEOTC (17%), English (12%), and then science (8%). The researchers expressed surprise at the low level of use in science, but discovered that "earth science is lightly featured in this area of the primary school curriculum" (p. 34). Reasons given by teachers for not using the resource, even though it was perceived to be valuable, related to the need to find space in an already overcrowded curriculum.

Curriculum Response to the Canterbury Earthquakes

In primary schools and early childhood centres, the curriculum response following the 2010–2011 earthquakes rarely focused on scientific understanding about the earthquakes. Instead, there was concern about its effects and how to look after oneself and others—the focus was on supporting students to overcome their trauma: "Pastoral care and wellbeing were the most important focus at the time of the immediate crisis and in the aftermath" (ERO, 2013, p. 1). In early childhood centres and primary schools, efforts involved developing a curriculum responsive to the children's emotional needs, for example, enabling them to play out or write stories about their experiences, often through drama, music, and art. This was especially so for young children. For example, Bateman and Danby (2013) emphasised the importance for preschool children of sharing memories as a way of recovering from disaster experiences. They described how one early childhood teacher's concern about supporting a child through sharing experiences allowed her to accept his working theory of the earthquake as "the dinosaurs were dancing" (p. 470) and "they were stomping really hard" (p. 472).

For older children, at a level where there is an explicit curriculum, that curriculum was likely to continue as usual. Based on interviews with principals, teachers, parents, and students in four primary schools across Canterbury in November 2012, O'Connor (2013) reported that

> Perhaps surprisingly, the teachers I spoke to said that although they spent time talking with children more, hugging and sharing tears with them, and playing games more, they said there was no change to actual curriculum content. The stories they read together, the content of their curriculum continued as if the quake had never happened. Teachers understood that returning to routine meant not addressing the issues that children faced on a daily basis in curricular work.... [T]heir curricular work was about preparing for the future rather than helping children make sense of their present. (p. 430)

Although O'Connor (2013) found no instances of teachers using a curricular means of looking at change and how people coped with the earthquakes and their aftermath, the ERO (2013) reported "examples of teachers using the earthquakes as a theme or topic in the curriculum" (p. 15). For example, in one primary school, children were helped to focus on identifying and naming their feelings by designing and making masks about the earthquake (Ormandy, 2014). Overall, however, the focus was on providing emotional support for children and their families.

In secondary schools there was a greater emphasis on providing curriculum continuity. Where attendance was difficult due to school closure or site-sharing, teachers maintained contact with their students and families via websites, intranet, Facebook, and blogs. For senior students the usual high-stakes assessment criteria were still in place. The ERO (2013) reported that "students across greater Christchurch achieved some of the best National Certificate of Educational Achievement (NCEA) results in New Zealand in 2011", and "the Chief Executive of the New Zealand Qualifications Authority (NZQA) stated that this was not the result of the special 'Earthquake Exemption' derived grades introduced for course endorsement for 2011 but a 'testament to the students, their teachers, principals and parents'" (p. 12).

In schools where the Canterbury earthquakes were most felt, the coping response in curriculum was therefore generally oriented towards maintaining the status quo of schooling. Although community involvement was strong, it was focused on pastoral care, emotional support, and helping schools resume their normal routine. Johnson and Ronan (2014) explored curricular responses to the earthquake in primary schools beyond the Canterbury region as part of an evaluation of "What's the Plan Stan?" simply because their data collection had been planned to occur in March–April 2011, just after the February earthquake. They analysed data collected from seven focus groups including 49 educators from 31 schools across New Zealand, but excluding Canterbury whose focus group was cancelled. Teachers reported that children generally wanted to talk about the earthquakes and this was an almost daily topic of conversation, particularly among younger children. Some interesting classroom discussions were had, some teachers did reading activities with relevant books, and others had children writing about the earthquake and their feelings. However, all of the schools had enrolled some students who had been displaced from the Christchurch area and teachers were hesitant about how to deal with the topic more formally. Many of the displaced children were traumatised by their earthquake experience. Often these children were helped by local students talking to them, asking questions, and listening to their stories. Most teachers were unsure how to interact with these children, many of whom did not wish to be reminded of their experiences. Nevertheless, teachers believed that "they would address children's reactions to the best of their ability after a disaster, even if they are not certain about the best approach" (p. 1082).

In schools geographically further afield, there was different scope for a curriculum response. Taylor and Moeed (2013a, 2013b) endeavoured to track the curricular response to the Canterbury earthquakes in New Zealand secondary schools. Writing in the context of disaster education, Taylor and Moeed (2013a) pointed out that, in terms of formal curriculum requirements, "[t]he alignment of national curriculum achievement objectives with NCEA [New Zealand's senior secondary school qualification] achievement standards make it clear that geographic learning is expected to integrate both the natural processes and the human dimensions to disasters such as earthquakes" (p. 59). They noted that general science teachers were advised that students were expected to demonstrate understanding of the formation of surface features. Teachers were also advised by the New Zealand Qualifications Authority (NZQA) "that students may be assessed on internal processes such as 'movement along fault lines, folding, faulting, and uplift' and/or 'land movement due to earthquakes'" (Taylor & Moeed, 2013a, p. 60). While this contains no overt reference to disaster education, Taylor and Moeed pointed out that the New Zealand curriculum encourages teachers to respond to particular interests of students, so teachers of geography, social studies, science, and other subjects may incorporate topics related to extreme events, such as earthquakes, should these occur.

In late 2010, following the September earthquake, Taylor and Moeed (2013a) distributed a questionnaire to secondary social science and science teachers to determine their initial curriculum response. Of the 53 teachers responding across New Zealand, not all had chosen to explicitly address the earthquake. In addition, senior teachers of geography and junior social studies (often taught by the same teachers) devoted more lesson time to the earthquake than science teachers. Teachers agreed that the importance of the earthquake to New Zealand and responding to student interest were the greatest facilitators to teaching the topic. The main barrier perceived by teachers not including any earthquake response was being locked into a curriculum and assessment time frame, particularly in senior classes. Most teachers agreed that the major outcomes of any curriculum response were knowledge related to the earthquake, possibly because the survey occurred towards the end of the school year at examination time. For senior school geography teachers who had some focus on attitudes and values, this focus decreased with distance from the Canterbury-Christchurch region.

After the February 2011 Christchurch earthquake, Taylor and Moeed (2013b) explored longer term curricular responses by interviewing seven science teachers, seven geography teachers, and two teachers who taught both subjects. These teachers self-selected from those who had responded to the earlier survey and were from seven schools in Auckland, Wellington, Dunedin, and Christchurch. Teachers chose a total of 32 articulate Year 11 students from their classes, and data were collected from both teachers and students via focus groups in November–December 2011 and March 2012. Again, knowledge outcomes were given prominence and again it was found that the curricular response was stronger in geography than science. Geography teachers used the earthquakes as a basis for case studies, and "[a]ll geography teachers, except one in Christchurch, indicated that the earthquake series had become part of their planned and taught curriculum" (p. 19). In contrast, only one science teacher said it had become a feature of the planned curriculum, and it was evident that responses to the earthquakes in science were a result of students' questions and interest. All science teachers agreed there was more time and scope to include the topic in the junior compared to the senior curriculum. Students from all schools distinguished the curriculum contribution from science to be more technical and from geography, more humanising. In addition, Taylor and Moeed found that "teacher capacity, distance from Christchurch, curriculum pragmatism, compassion, and perspectives of learning were all given as reasons by teachers for omitting the Canterbury-Christchurch earthquakes from their curriculum-making" (p. 22). Two science teachers used the claims of a local forecaster to predict the earthquake as "a useful opportunity to explore an authentic science-in-the-media controversy so that students think critically about the nature and weight of scientific evidence" (p. 20).

This last example was one of very few forays into the informal sector (in this case, popular media) that we found in our search for curriculum response to the Canterbury earthquakes. Reflecting on their findings about children's understanding of natural hazards 10 years previously, Johnston et al. (2014) noted that numerous researchers were currently exploring the impact on children of the Canterbury earthquakes, particularly children's role in creating "the narrative of the earthquake, and the role of schools in the response and recovery process" (p. 66). No doubt research results not yet available will reveal more comprehensively the ways in which teachers adapted their curriculum—and interacted with out-of-school communities—in response to the disaster. It is likely these responses will reveal some integrated approaches, particularly at the junior school level, and particularly relating to disaster awareness education.

Discussion

We began this chapter by arguing that effective science education needs to reflect our (inter)connected world, and to do so the school curriculum must not only educate students prior to significant science-related events that arise in the community outside of school, but also respond and adapt to the post-event community and environment. To illuminate this argument, we explored the content and nature of the curriculum and the educational, and particularly the curricular, responses to earthquakes as a special case of a socioscientific issue that has deep emotional, social, economic, and political consequences. We saw how, following the Canterbury earthquakes in New Zealand, the immediate—and appropriate—educational responses were designed to support students to work through their trauma. Under difficult circumstances, teachers sought to find the appropriate balance between coping with immediate emotional responses and carrying on with students', teachers', and families' lives. In early childhood and primary classrooms, earthquakes sometimes became a curriculum theme that allowed children to play out their experiences, but for older children, the curriculum mostly carried on as usual—albeit in different locations for some. Based on his discussions with teachers in four Christchurch primary schools, O'Connor (2013) stated: "It can be argued that a curriculum which is futures-focused, driven by literacy and numeracy demands, has relatively little to say to teachers in a time of crisis, except to carry on as if nothing has happened" (p. 430). In secondary schools, continuity of the discipline-based curriculum was paramount, with considerable efforts made to support students' learning with the use of information and communication technologies and collaboration between learning institutions (ERO, 2013). We saw that in times of local stress, the default response was to return to the status quo, to "normalise" school life by getting back to the curriculum routines of study, subjects, and (particularly for upper secondary students) preparation for assessment.

This response is entirely understandable. It is also understandable that teachers tried to revert to "business as usual" as a means of coping both for their students and for themselves. Earthquakes (and similar sudden disasters) create enormous disruption, anxiety, fear, and stress, even if there is no loss of life. The National Child Traumatic Stress Network (NCTSN, n.d.) in the US published "Teacher Guidelines for Helping Students after an Earthquake" which describes common reactions and suggestions about what teachers can do to assist their students to cope. First, teachers are advised to take care of themselves, thus creating the emotional space to take care of others. Very few suggestions refer to curriculum, but all refer to assisting students to deal with ongoing, stress-related responses.

In Canterbury, community facilities responded to assist schools, teachers, and students during the recovery process. *Science Alive!*, the local science centre, lost its building in the Christchurch earthquake, but within 3 weeks was back in action, supporting schools with travelling science curriculum-based resources that included hands-on learning with pre- and post- activities designed to promote understanding of the content. Teachers were able to select a topic and structure lessons around it in their own classrooms. Significantly, *Science Alive!* staff worked with the National Mental Health Foundation to develop programmes based on the Mindball Game (www.sciencealive.co.nz/mindball) designed to assist students become more relaxed and focused and learn techniques for doing this. *Science Alive!* staff also presented shows and exhibits taken into community venues such as libraries and market places. All of these programmes continue as *Science Alive!* works to promote a sense of normalcy in the community (Neville Petrie, personal communication, 22 September 2015).

The Canterbury earthquakes exemplified the importance of community links, especially between affected schools and their students' parents. In areas beyond Canterbury there was relatively limited curriculum response, even though the curriculum gave teachers agency to adapt their teaching to students' needs and interests. Taylor and Moeed's (2013a, 2013b) analyses of teachers' curricular responses to the earthquakes found that in science, there tended to be greater response at the junior secondary level where teachers had more flexibility to respond to students' interest, but in social studies there was greater curriculum response in senior secondary geography.

Taylor and Moeed (2013b) also referred to "the perennial issue of tension between permissive curriculum and constraining assessment in the senior secondary school" (p. 24). Unless teachers have ways to assess students' broader learning in integrated, community-linked programmes, they will fall back on assessing content knowledge, as did some of the teachers in Taylor and Moeed's studies. The NZQA assessment of students' understanding of extreme events in geography enabled teachers to include it in the curriculum. In senior science, where study of physics, chemistry, and biology was prioritised over earth science, it was more difficult to include even local earthquake events in the curriculum. In both Taylor and Moeed's research and the evaluation of "What's the Plan Stan" (MCDEM, 2012), teachers referred to the pressures and inflexibility caused by a crowded curriculum, an additional reason to maintain the status quo.

What we have seen in terms of school responses in the immediate aftermath of the Canterbury earthquakes is the turning of attention to the emotional support of students, teachers, and families, and efforts in combination with the community to return to normal schooling and curriculum stability in students' lives. We empathise with and respect the responses made by school personnel to place the emotional needs of their students before curriculum considerations. Also, it is not surprising that teachers, especially those of senior classes, returned as soon as possible to their standard curriculum practices, given the pervasive influence of the grammar of schooling, in terms of discipline-focus and pressures of assessment. Further, as Taylor and Moeed (2013b) suggested, science teachers are not well equipped to teach outside of their traditional content knowledge base and this brings into question their ability to deliver important disaster preparedness education. While some early research (Finnis et al., 2004) suggested that more disaster awareness was required, it is worth noting that there was not a single serious personal injury in a New Zealand school during the earthquakes.

What other curricular responses might be possible? Earthquakes, like other natural disasters, are events that may occur locally but have global repercussions. All kinds of science are used to document them, learn how they occur, and try to predict their occurrence and magnitude. Earthquakes are not merely events, they require recovery and reconstruction, and that includes a range of different sciences and other related disciplines, such as mathematics and engineering, and also medicine, agriculture, and social sciences. Earthquakes are also about people, and that makes major earthquake events of significant international interest and aid agencies are invariably required to assist, particularly in the early stages of recovery.

It is clear that earthquakes and their aftermaths involve much more than science, inviting an interdisciplinary/multidisciplinary approach to curriculum. There are numerous resources in schools and in the community to support such an approach. While school textbooks usually focus on generic explanations for earthquakes relating to plate tectonics, other kinds of information are readily available online in both text and visual media, as well as in natural history museums and science centres. In addition, in disaster prone areas with robust educational systems there are invariably excellent supplementary curriculum resources available on the topic of disaster preparedness. We can also find evocative images readily available on the internet, or activities available in connection with real-time earthquake-monitoring websites. We would argue that these along with many other sources of information make earthquakes potentially one of the most resource-rich topics in the curriculum.

Because they are socioscientific issues, disasters such as earthquakes provide powerful opportunities to explore the nature of science. For example, the monitoring of earthquakes and their effects illustrates knowledge building in science and other disciplinary areas as seismologists try to make sense of the resulting data. New theories on liquefaction and recognition of the "slap-down effect" based on information derived from the New Zealand Canterbury earthquakes demonstrate that science is cumulative and based on evidence. In fact, resources developed to support science learning following the Canterbury earthquakes focus particularly on these aspects as a means of developing understanding of the "Use Evidence" capability related to the Nature of Science strand of the New Zealand Curriculum (Simpson, 2014).

But there are also aspects relating to disaster education, before and after the earthquake, that address the human impact of earthquakes, as well as possible effects on school curricula in science and other disciplines. Many of these can benefit from a cross-curricular approach with effective community links. Although Taylor and Moeed (2013a, 2013b) made no specific mention of integration across the curriculum, they reported that

... some notable teacher responses to the Canterbury earthquake series indicated manoeu-
vring into spaces of scientific literacy and critical geography. This was particularly the case
when teachers seized on the media commentary about [a local forecaster's] predictions of
earthquakes based on phases of the moon. (2013b, p. 24)

The issue of predicting earthquakes is an interesting one for students to pursue
from an integrated perspective because it has significant social implications as well
as a strong science-related disciplinary base. The likelihood of earthquakes in particular areas is well-known, but the exact location or timing is not possible to predict. For example, in terms of location, it is notable that the 2011 Japan earthquake
and the Boxing Day 2004 Sumatra—Banda Aceh earthquake both occurred in parts
of the earth crust that geologists had considered were relatively stable (Goldfinger,
Ikeda, Yeats, & Ren, 2013), and the Canterbury earthquakes occurred on an
unknown fault line. With regard to timing, historically, Nepal has a large earthquake
about every 80 years and the previous major earthquake was in 1934. Nepal com-
memorated the 80th anniversary of that earthquake in January 2014 on "Earthquake
Day". At the time, the editor of Nepal's New Spotlight Magazine, wrote "As Nepal
is celebrating the Earthquake Day, remembering the horror of the 1934 earthquake,
the time has come for policymakers, civil society and people to work and minimize
the damage in life and property if an earthquake hits" (Poudel, 2014, para 1). He
went on to review Nepal's major earthquakes over the last eight centuries and
emphasised the expectation of another earthquake soon. He described recent efforts
in Nepal's risk management and mitigation programmes, many with international
partners, including school safety education. However, in light of the slowness to
tighten building codes and educate the population about disaster preparedness,
Poudel warned that "the looming earthquake is likely to be devastating for the people and their life and property" (final para). Sadly, he was correct. Just 15 months
later, Nepal was devastated by the April 2015 earthquake.

From another perspective, the prosecution of six Italian scientists and a public
official for failing to predict the L'Aquila earthquake (Johnston, 2012) is an interesting case study involving both science and social studies. Despite the defence lawyers protesting that major earthquakes could not be predicted, prosecution lawyers
argued that the defendants were falsely reassuring that an earthquake would not
occur. The trial resulted in a conviction for manslaughter and a six-year jail term for
the seven defendants. This verdict was subsequently quashed on appeal for the six
scientists, and the public official had his sentence reduced to 2 years (Squires,
2014). The potential ramifications for science if scientists can be prosecuted for failing to predict what cannot be known could lead to robust discussion among senior
students about the fallibility of science, its reliance on evidence, replication, and so
on.

Another example of the interweaving of socioscientific issues with science and
its curriculum is the significance of communication. When a disaster occurs, it is
essential to communicate to the local community what is happening and what needs
to be done for its safety. The EEW system in Japan gave most of its citizens enough
time to take protective measures. In New Zealand, following the Canterbury earthquakes, many schools communicated with their parent body via social media to

inform them of procedures to collect children after the disaster and then to maintain contact through the gradual resumption of schooling (ERO, 2013). For communities beyond the disaster zone, the pervasiveness of electronic media connects the world with the event. Much of the information coming from Nepal in the wake of the earthquake, aside from official channels such as the UN News Centre, came from western tourists, nearly all of whom were carrying cameras. Within 6 weeks of the earthquake on 25 April 2015, the Smithsonian Channel was broadcasting its 46-minute documentary of the "Nepal Quake: Terror on Everest", derived essentially from the visuals recorded as the disaster unfolded by mountaineers in camps on Mt. Everest, supplemented by vision from tourists with smart phones. Other footage in Kathmandu came from two students who had been filming for their YouTube channel. The programme included dramatic footage of the avalanche that destroyed Base Camp and some of the landslides that covered villages in the mountains following the large aftershock on 12 May, together with stories from survivors and interviews with three geologists about the location, nature, and inevitability of an earthquake given Nepal's geographic location. The documentary was posted to YouTube on 10 June 2015.

Conclusions

In this chapter, we illustrated the need for greater complementarity between the traditional, unidisciplinary science education students are likely to experience in school and the wide variety of interdisciplinary resources and opportunities offered for science learning available outside of school. Using the Canterbury earthquakes as a case study of a special kind of socioscientific issue, we explored how school curricula responded to a significant, science-related community event. The effects of earthquakes and other natural disasters are long lasting and they involve not just science, but every aspect of people's lives. Disaster awareness programmes are an important part of education in disaster prone countries, and well suited to an interdisciplinary cross-curricular approach that, by connecting with the community, can merge the in-school and out-of-school sectors of science education. Although natural disasters may be a particular kind of socioscientific issue, we suggest that an interdisciplinary cross-curricular approach with links to the community is a necessary precursor to dealing effectively with any socioscientific issue (see Chapter "Pregnant Pauses: Science Museums, Schools and a Controversial Exhibiton" by Erminia Pedretti and Ana Nava-Iannini, Chapter "Reinvigorating Primary School science Through School-Community Partnerships" by Kathy Smith and Angela Fitzgerald, and Chapter "The Challenges and Opportunities for Embracing Complex Socio-scientific Issues as Important in Learning Science: The Murray-Darling River Basin as an Example" by Peter Fensham and Jasper Montana).

Nielsen (2013) argued that in making decisions about socioscientific issues, scientific evidence should not have a privileged role. Instead, it is the quality of a student's deliberation and use of evidence about the issue that should be the important consideration. Nielsen stated that "socioscientific decision making should be conceptualised as a deliberation about what to do about topical societal issues that relate to science" (p. 38). Deciding what to do introduces the need to consider values and ethical principles as well as scientific evidence. Nielsen refers to teachers' potential discomfort in dealing with socioscientific issues that they may perceive as involving political and ethical dimensions at the expense of scientific rigour. Indeed, teaching about socioscientific issues does present a professional challenge to many teachers (Rennie, 2011). However, students' lives outside of school are not immersed in scientific rigour; they are immersed in the values, politics and immediacy of the real world. This is particularly true in the case of natural disasters such as earthquakes.

In closing, we would like to honour the work of New Zealand schools and their communities in responding to the Canterbury earthquakes. As much as anything, the New Zealand experience illustrates the complexity and intricacy of working in the shadow and aftermath of a natural disaster. These events were bigger in the lives of those involved than those of us who were not there can imagine. Some may argue that an earthquake is almost too big for schools to deal with, that the private and personal affect should be separated from the public role of schools to deliver the mandated curriculum. Our own position is that natural disasters are of such proximity and consequence that they warrant a direct curriculum response. Such a response, by its very nature, will be both flexible and integrated across disciplines and with the community, and will draw on resources near and far. There are no easy answers here, for the responses will vary from community to community and with different natural disasters. But the aim is to empower students, their families, and their teachers, to help them to understand, cope with, and respond to events so central to their lives.

References

Alexander, D. (2012, August). A tale of three cities and three earthquake disasters. *Tafter Journal*, No. 50. Retrieved from http://www.tafterjournal.it/2012/08/01/ataleofthreecitiesandthreeearthquakedisasters/

Bateman, A., & Danby, S. (2013). Recovering from the earthquake. *Disaster Prevention and Management, 22*(5), 467–479. https://doi.org/10.1108/DPM-10-2013-0177.

Education Review Office. (2013). *Stories of resilience and innovation in schools and early childhood services.* Canterbury earthquakes: 2010–2012. Wellington, New Zealand: Author.

Finnis, K., Standring, S., Johnston, D., & Ronan, K. (2004). Children's understanding of natural hazards in Christchurch, New Zealand. *The Australian Journal of Emergency Management, 19*(2), 11–20.

Goldfinger, C., Ikeda, Y., Yeats, R. S., & Ren, J. (2013). Superquakes and supercycles. *Seismological Research Letters, 84*(1), 24–32.

Japan Meteorological Association (JMA). (2013, October). *Lessons learned from the tsunami disaster caused by the 2011 great East Japan earthquake and improvements in JMA's tsunami warning system.* Retrieved from http://www.data.jma.go.jp/svd/eqev/data/en/tsunami/LessonsLearned_Improvements_brochure.pdf

Johnson, V. A., & Ronan, K. R. (2014). Classroom responses of New Zealand school teachers following the 2011 Christchurch earthquake. *Natural Hazards, 72*(2), 1075–1092. https://doi.org/10.1007/s11069-014-1053-3.

Johnston, A. (2012, October 22). L'Aquila quake: Italy scientists guilty of manslaughter. *BBC News*. Retrieved from http://www.bbc.com/news/world-europe-20025626

Johnston, D., Ronan, K., & Standring, S. (2014). Children's understanding of natural hazards in Christchurch: Reflecting on a 2003 study. *The Australian Journal of Emergency Management, 29*(1), 66.

Ministry of Civil Defence & Emergency Management (MCDEM). (2012). *Report of the 2012 "What's the Plan Stan?" survey of New Zealand primary schools*. Auckland, New Zealand: Author.

Murakami, S. (2014, March). How broadcasters used the Internet: Simulcasting at the time of the Great East Japan earthquake. *IEEE Communications Magazine, 52*(3), 51–55.

Mutch, C. (2014). The role of schools in disaster preparedness, response and recovery: What can we learn from the literature? *Pastoral Care in Education: An International Journal of Personal, Social and Emotional Development, 32*(1), 5–22. https://doi.org/10.1080/02643944.2014.880123.

Mutch, C., & Marlowe, J. (2013). Lessons from disaster: the power and place of story. *Disaster Prevention and Management, 22*(5), 385–394. https://doi.org/10.1108/DPM-10-2013-0172.

National Child Traumatic Stress Network (NCTSN). (n.d.). *Teacher guidelines for helping students after an earthquake*. Retrieved from http://www.nctsn.org/sites/default/files/assets/pdfs/Teachers_Talk_to_Students_about_Earthquake.pdf

Nielsen, J. A. (2013). Delusions about evidence: On why scientific evidence should not be the main concern in socioscientific decision making. *Canadian Journal of Science, Mathematics and Technology Education, 13*(4), 373–385.

NZ Ministry of Education. (2007). *The New Zealand curriculum*. Wellington, New Zealand: Learning Media.

NZ Ministry of Education. (2009). *EOTC guidelines: Bringing the curriculum alive*. Wellington, New Zealand: Author.

O'Connor, P. (2013). Pedagogy of love and care: Shaken schools respond. *Disaster Prevention and Management, 22*(5), 425–433.

Ormandy, S. (2014). Wellbeing and the curriculum: One school's story post-earthquake. *Teachers and Curriculum, 14*, 3–11.

Poudel, K. (2014, January 14). Kathmandu Valley quake: Real risk. *New Spotlight News Magazine, 7*(14). Retrieved http://www.spotlightnepal.com/News/Article/KATHMANDU-VALLEY-QUAKE-Real-Risk

Poudel, K. (2015, May 21). Deadly Gorkha earth quake boon in disguise. *New Spotlight News Magazine, 8*(21). Retrieved http://www.spotlightnepal.com/News/Article/Earthquake-Nepal-Devastation-Keshab-Poudel

Rennie, L., Venville, G., & Wallace, J. (2012). *Knowledge that counts in a global community: Exploring the contribution of integrated curriculum*. London, UK: Routledge.

Rennie, L. J. (2011). Blurring the boundary between the classroom and the community: Challenges for teachers' professional learning. In D. Corrigan, J. Dillon, & R. Gunstone (Eds.), *The professional knowledge base of science teaching* (pp. 13–29). Dordrecht, The Netherlands: Springer.

Schwarz, E., & Stolow, D. (2006). Twenty-first century learning in afterschool. *New Directions for Youth Development, 2006*(110), 81–99.

Simpson, P. (2014). Learning from the Christchurch earthquakes. Teacher support material for "Learning from the Christchurch Earthquakes". *Connected, Level 4*. Retrieved from www.connected.tki.org.nz

Squires, N. (2014, November 10). *Italian scientists cleared of failing to predict L'Aquila earthquake*. Telegraph Media Group. Retrieved from http://www.telegraph.co.uk/news/worldnews/europe/italy/11221825/Italian-scientists-cleared-of-failing-to-predict-LAquila-earthquake.html

Stocklmayer, S. M., Rennie, L. J., & Gilbert, J. K. (2010). The roles of the formal and informal sectors in the provision of effective science education. *Studies in Science Education, 46*, 1–44.

Taylor, M., & Moeed, A. (2013a). The 2010 Canterbury earthquake: Curriculum shockwaves. *International Research in Geographical and Environmental Education, 22*(1), 57–70. https://doi.org/10.1080/10382046.2012.759693.

Taylor, M., & Moeed, A. (2013b). Curriculum shockwaves? Geography, science, and the Canterbury earthquakes. *Curriculum Matters, 9*, 8–28.

Tylack, D., & Tobin, W. (1994). The grammar of schooling: Why has it been so hard to change? *American Educational Research Journal, 31*(3), 453–480.

UN News Centre. (2015a, June 5). *Nepal's heritage sites on shaky ground after devastating quake.* Retrieved from http://www.un.org/apps/news/story.asp?NewsID=51065#.VXunlPnvPIU

UN News Centre. (2015b, June 9). *Ban welcomes 'milestone' agreement on new Nepal constitution.* Retrieved from http://www.un.org/apps/news/story.asp?NewsID=51102#.VXumm_nvPIU

Vervaeck, A., & Daniell, J. (2011, April 22). *2 examples of how technology made the difference during the Tohoku earthquake and tsunami.* Retrieved from http://earthquake-report.com/2011/04/22/2-examples-of-how-technology-made-the-difference-during-the-tohoku-earthquake-and-tsunami/

World Health Organization (WHO). (2015, May 13). *Emergency preparedness pays off as Kathmandu hospitals respond to earthquakes.* News release. Retrieved from http://www.who.int/mediacentre/news/releases/2015/nepal-second-quake/en/

Yamasaki, E. (2012). What we can learn from Japan's earthquake early warning system. *Momentum, 1*(1), Article 2. Retrieved from http://repository.upenn.edu/cgi/viewcontent.cgi?article=1022&context=momentum

The Challenges and Opportunities for Embracing Complex Socio-scientific Issues As Important in Learning Science: The Murray-Darling River Basin As an Example

Peter J. Fensham and Jasper Montana

Abstract Socio-scientific issues present a great challenge to science educators that are charged with equipping students—as future adult citizens—with the knowledge, skills and attitudes to understand and respond to them. These issues, such as climate change and over-exploitation of resources, are increasingly prominent in our lives. Complex socio-scientific issues are often defined by an interrelated set of smaller issues, they can have vast social impacts and their scientific basis is often uncertain or contested. The increasing global conflict around water, in particular in rivers that flow across territorial or national boundaries, is a notable example of one of these issues.

In Australia, the management of the Murray-Darling River Basin, which underpins a large part of the nation's agricultural economy, became the focus of intense public debate in all forms of the media between 2010 and 2012. At the same time, a new national curriculum for school science was being developed. In this chapter, we use the Murray-Darling controversy as a context to investigate how this science curriculum might facilitate teaching and learning of socio-scientific issues (SSIs) by considering this SSI. We adopt the analytical tools of frame theory and boundary work to assess:

(i) the role of science in the controversy surrounding this SSI;
(ii) the strengths in the science curriculum to make a contribution to understanding the science involved; and

P. J. Fensham (✉)
Faculty of Education, Monash University, Clayton, VIC, Australia
e-mail: peter.fensham@monash.edu

J. Montana
Department of Politics, University of Sheffield, Sheffield, UK

© Springer International Publishing AG, part of Springer Nature 2018
D. Corrigan et al. (eds.), *Navigating the Changing Landscape of Formal and Informal Science Learning Opportunities*,
https://doi.org/10.1007/978-3-319-89761-5_8

(iii) the lessons that can be drawn from the Murray-Darling controversy about how the science curriculum might better equip teachers and students to tackle such complex SSIs.

Keywords Socio-scientific issues · Science curriculum · Framing · Boundary work · Water management · Science controversy

Introduction

Ever since the official endorsement of 'Science for All' in the 1980s (Fensham, 1985), authorities responsible for science education have acknowledged that school science has a definitive role in equipping all students, as future adult citizens, with the knowledge, skills and attitudes needed to participate in the ways that science and technology (S&T) influence society.

As a first response to 'Science for All', there was a spate of interest in introducing real world S&T contexts into the science classroom. These innovations used technological applications of science to link science and society and collectively they became the Science–Technology–Society (STS) movement (Solomon & Aikenhead, 1994). Unfortunately, they made little impact on official science curricula because wider reforms of schooling were occurring, which included establishing Technology as a new subject area for a number of existing 'make and design' subjects. This sense of 'Technology' was very different from the 'applications of science' meaning it had in the STS movement. Nevertheless, 'making decisions about S&T issues' persisted as a goal of school science and became increasingly common among its list of intended outcomes (Aikenhead, 1992; Kortland, 1996; Ratcliffe, 1997)—although there is little evidence from international curriculum-based studies since 1994 to indicate that 'making decisions' has become a serious part of the mainstream school science curriculum (Thomson, Hillman, & Wernert, 2012).

How this learning outcome can be taught in school science has, nevertheless, been a major twenty-first century interest for science education research. Some of these studies have been concerned with the scientific processes that are components of decision-making. Others have gone further, engaging the teacher and students with an actual or précised account of a real world science and technology issue in order to make decisions about it. Because these issues have societal dimensions as well as scientific ones, they are referred to as socio-scientific issues (SSIs).

SSIs vary greatly in the science that is involved and in their societal impact. Some involve relatively simple knowledge of science and its application while others require both disciplinary and interdisciplinary scientific knowledge, as well as a substantial appreciation of the Nature of Science. In relation to the societal influence of SSIs, this can range from a localised group of citizens to large sections of a national society and beyond to the international community. At the latter end of these two spectra, where the science knowledge and the ramifications on society are both very broad and significant, the issues can be described as 'complex'.

In this chapter we discuss the challenges and the opportunities that complex SSIs present to school science education. As an example, we use the case of the Murray-Darling River Basin in Australia to provide a set of scientific reference points against which Australia's new national science curriculum is judged for the opportunities it provides for teaching such a complex SSI. In doing so, we draw out in detail the science of this SSI and use insights from the sociology of science to examine how scientific evidence and expert authority are sometimes challenged by public scrutiny.

We conclude by reflecting on both the opportunities and challenges presented by the inclusion of complex SSIs in school science education. For example, SSIs challenge the traditional notions of students as intellectually independent learners in school science education, and suggest that a goal of *intellectual dependence* is also needed. Furthermore, we argue that some key common features of the science in SSIs will require a willingness among science teachers to embrace new pedagogies in their classrooms.

Complex Socio-scientific Issues

A number of more complex SSIs were identified in the beginning of the twenty-first century as Grand Challenges and Opportunities (AAAS, 2006; National Research Council, 2001). Decision-making about them falls into Rittel and Webber's (1973) category of 'wicked problems' because they are not single issues but are made up of a set of inter-related issues. Often these issues require high-stakes decisions to be made urgently, when the scientific and societal aspects are still uncertain and incomplete, and the related values are in dispute. Funtowicz and Ravetz (1993) invoked the term 'post- normal science' to describe the type of knowledge required for this decision-making. Once the SSI is presented in the public arena the diversity of stakeholders often expands, drawing into question the type of evidence that is considered relevant and who should be considered an expert.

Complex SSIs present a great challenge to school science education, because the science involved is invariably beyond what is included in the curriculum. Nevertheless, because they have significance for so many citizens and future generations they cannot be ignored. Some current ones are issues associated with climate change, food production, biodiversity, and the over exploitation of natural resources. Prominent among the last of these is the issue of access to water, which crosses territorial or national boundaries and is a matter of increasingly intense debate and diplomacy (Poff et al., 2003; Sullivan, 2014).

The Murray-Darling River Basin is an Australian example of an *access to water* issue. The Murray-Darling is Australia's largest river system, spanning over one million kilometres and passing through four of Australia's six states, each of which historically held the rights and responsibilities for the use of its water. Environmental concern and the intervention of the national government have recently led to the development of a management strategy for the river system. This intervention mobilised a diverse range of stakeholders. It also incited conflict over the social needs for water extraction and the environmental requirements for the rivers basin's functioning ecosystems.

The attempt to manage the Murray-Darling River Basin highlighted the scientific, economic, cultural and political considerations that are involved in all complex SSIs. In particular, it demonstrated how scientific evidence and authority is defined and challenged in public controversy.

The Renewal of Interest in S&T Issues for School Science

In the early 1990s, as the STS movement was petering out in relation to school science, a few science educators began to conduct case studies of small groups of citizens who had a 'need-to-know' about a local S&T issue affecting them (Irwin & Wynne, 1996; Layton, Jenkins, Macgill, & Davey, 1993). Even when the science involved was relatively simple, they found it needed more direct translation and its trustworthiness had to be explained. When decisions were made the science was still weighed against a range of other information. A review of 31 studies by Ryder (2003) confirmed the need for understandable scientific information but also for some appreciation of science's procedures for achieving knowledge, including how data are evaluated and used as evidence.

To an extent inspired by these studies in the public arena, there was a renewal of interest among science educators in 'making decisions' about S&T issues in school science. For example, Driver, Newton, and Osborne (2000) introduced into science classrooms scientific argumentation as it relates to decision making in both scientific contexts and in S&T issues. This extended the meaning of the 'Nature of Science' beyond the procedural sense of inquiry it had hitherto had in school science (Bell & Lederman, 2003). Others (e.g., Kolstø (2001) in Norway, and Zeidler, Sadler, Simmons, and Howes (2005) in the USA) took these studies a step further by introducing school students to a précised outline of an SSI issue and engaging them with its particular science, its scientific procedures, and its social implications. Of course, the content knowledge of science varies from issue to issue, but it can be taught in the classroom if it is relatively simple or it may have already been covered in science curriculum. However, what is common to SSIs, and to making decisions about them, is this more extended notion of the Nature of Science: an opportunity and a challenge for any science curriculum, as we discuss later in the chapter.

The studies of SSI science teaching have led to considerable debate among science educators about the extent to which the non-science aspects of an SSI should be included in science teaching. Levinson (2006, 2010) provided frameworks (See chapter "Pregnant Pauses: Science Museums, Schools and a Controversial Exhibition", this volume) for teaching both the scientific and social aspects of SSIs that call for a dialogic and democratic style of pedagogy that is very different from the transmissive and authoritative discourse that is so commonly used in science classrooms. Zeidler and Sadler (2008) have stressed the importance of making moral and ethical aspects explicit. Bencze and Carter (2011) argued for a more radical extension of Hodson's (2003) call for students to engage in socio-political action about the issues. On the other hand, Hodson, Bencze, Elshof, Pedretti, and Nyhof-Young (2002) and Levinson (2004) both provide cautionary evidence from science teachers and students against extending the boundaries of science education too far.

We do not contest the importance of the non-science components of SSIs. It will be evident that they were very important in deciding the final management plan for the Murray-Darling. However, as far as the mainstream science classroom is concerned, we assume that a lay acknowledgement and open discussion and debate about them is what ought to be expected of teachers and their students if SSI teaching is to be established in mainstream science teaching. Accordingly, in our analysis of this example of a complex SSI, our focus is on its science content and scientific procedures and how these were played out in the public debate and in the political resolution of the issue.

The findings of the analysis then become a set of reference points against which Australia's new national science curriculum is judged for the opportunities it provides for teaching this SSI's science content and procedural knowledge, and how it is challenged to better contribute to such decision-making.

Socio-scientific Issues and Public Deliberation

The deliberation of complex SSIs in public arenas can frequently erupt into controversies that provoke fervent and widespread disagreement. Over recent decades, sociologists of science have made extensive use of these controversies to produce important insights about the relationship between science and society (for a foundational example see Nelkin, 1979). In practice, these studies reveal that the closure of a controversy is often not found in the traditional domain of science, but is the result of negotiation across a range of social spheres.

One key determinant of how a controversy finds closure is through the way in which the problem is framed. Developed in early work by Goffman (1974) as mental structures that facilitate our basic interactions with the world, frames have more recently been applied to controversy studies as "underlying structures of belief, perception, and appreciation" that determine our policy positions (Rein & Schön, 1994, p. 23). Framing can be used to understand how different interpretations emerge of what a problem is, what evidence or expertise is relevant, and how a problem should be resolved. In particular, it has been shown that SSIs that are treated as purely 'scientific' or 'regulatory' often become unravelled as other publics seek to make sense of, and respond to them. Bonneuil, Joly, and Marris (2008), for example, examined how a scientific framing of research into genetically modified crops in France was rapidly challenged by publics more concerned with questions of who should benefit from the technological development. As such, identifying and adapting to emergent framing of SSIs becomes an important part of understanding and responding to them. As demonstrated by the case of the Murray-Darling, competing frames can limit productive dialogue in resolving SSIs.

The application of frame theory in science education is still in development but it has been used in recent studies of the transfer of learning in science (Engle, 2006; Patchen & Smithenry, 2013). It also has rich potential for the new emphasis on communication in science that science curricula are now including (see Gilbert & Stocklmayer, 2013; and chapter "Communicating Science" by Sue Stocklmayer).

In addition to exposing SSIs to competing frames, public deliberation can also throw into question the boundaries of science itself. As science is increasingly called upon to respond to pressing societal concerns, the traditional demarcation of science as distinct from other forms of knowledge making can become challenged (Jasanoff, 1987). The strategic efforts used to maintain, or modify, the demarcation between science and non-science is called 'boundary work' (Gieryn, 1983). The boundaries between science and other knowledge sources become particularly important in the resolution of SSIs. As outlined above, complex SSIs are often reliant on forms of post-normal science and, amongst the emergent groups of spokespersons, claims to scientific expertise can easily become contested. In these volatile spaces, the pre-negotiated boundaries of science and politics can break down, making choices about who to trust as an expert ever more difficult (Collins, 2009).

In science education, boundaries are drawn as to what knowledge is included in the science curriculum, in the choice by teachers of what contexts to bring into the science classroom, and how they choose to teach about them. School roles, for example of 'science teacher', 'biology teacher' or 'physics teacher', can also reinforce boundaries. In wider society, beyond the school, when complex SSIs are involved, delineations break down, and who represents science is much less clear. In the example of the Murray-Darling River Basin, which we now describe, the debate surrounding its management illustrates this blurring of boundaries.

The Murray-Darling River Basin

As with many large-scale river systems worldwide, the Murray-Darling River Basin is a central feature of the national agricultural economy and has played an important part in the lives of indigenous people for thousands of years. The large-scale production of wool, cotton, and food, accounting for 40% of the national agriculture income, underpins the economies of hundreds of communities. However, it has become increasingly recognised that these extraction industries are now leaving too little water to sustain a healthy river system and threatening its long-term ability to support extractive uses into the future.

For over 100 years, the four state and the single territory governments located within the Basin have determined water allocation from the river. However, ongoing conflict between them has led to the need for a nationally-coordinated water management plan that would achieve more equitable access to water resources, while balancing the ecological health of the river and its associated ecosystems. By the start of the twenty-first century, the management of the Basin had become a hugely divisive environmental issue, the resolution of which had become increasingly politically urgent.

An initial attempt to introduce a national management plan was made in 2003, but increasing conflicts over how best to proceed scuppered any progress (see Crase, Dollery, & Wallis, 2005). In 2007, the Australian Government passed the Commonwealth *Water Act 2007,* which was intended to facilitate the reduction of water extraction to environmentally sustainable levels, while optimising associated social and economic returns. The Act established a small independent and expert body, the Murray-Darling Basin Authority (MDBA), to develop an implementation plan. The Authority was responsible for setting out evidence and providing a set of recommended policy options to the newly enacted Murray-Darling River Basin Ministerial Council, comprised of the national Minister for Water and a minister from each of the four state governments involved.

In 2010 the Authority published an initial *Guide to the Proposed Murray-Darling Basin Plan,* which set out clear environmental targets and recommendations for an annual allocation of water to be returned to the river (in Gigalitres per year, GL/y) for environmental flows (MDBA, 2010).

These targets took into account a range of aspects, including:

(i) the legal rights of indigenous persons to hunt, gather and fish in these inland waters

(ii) the conservation of the biodiversity of the rivers' ecosystems—the natural habitats of many species of flora and fauna are now so degraded that 16 of its 80 mammals and 17 species of fish are endangered

(iii) the protection of the Ramsar-declared wetlands and the internationally listed water-dependent ecosystems used by migratory bird species

(iv) the connectivity of the rivers and the flood plains

(v) the prevention of salination of the river flood plains, which threatens food webs that sustain water dependent ecosystems

(vi) the periodical opening of the mouth of the Murray River with sufficient flows through the Coorong, the narrow estuary from the river mouth to the sea

(vii) the relation between ground water and the rivers that maintains the quality of the water to sustain the dependent ecosystems

(viii) the system's resilience to climate change and to drought

(ix) threats from anthropogenic-related impacts (e.g., introduced species)

Most of these key targets were underpinned by scientific data and findings that had been integrated into the hydrological and environmental models used by the Authority. These scientific data and findings had varying degrees of certainty and so were assigned three levels of confidence: high (uncontested, peer reviewed), medium (data available, but not yet peer reviewed), and low (limited or emerging, needing more study). A majority of the science information had medium-level confidence and came from government-initiated studies and reports that had not been peer reviewed. The *Guide* specified that an allocation of between 3000 and 7600 GL of water should be returned to the river each year, with the two figures representing the low-level and high-level certainty of whether the targets identified in the *Water Act 2007* would be met. This volume of water to be allocated to the river became centrally important to the public debate around the issue.

The public and state government responses to the water allocations in the *Guide* were so critical that the national government reconstituted the Authority under a new chairman and a new executive officer. This new body presented an updated *Draft Basin Plan* in 2011 with an amended allocation of 2750 GL per year (MDBA, 2011a, 2011b). In contrast to the *Guide*, the uncertainties in the science were not explicitly addressed in the *Draft Plan* and the recommended allocation was not directly argued for on scientific grounds. Instead, the MDBA asked the national scientific body, the CSIRO, to convene an international team of scientific experts to review the quality of the scientific knowledge and procedures used in its development. The CSIRO (2011) review confirmed that sufficient science knowledge was available for the Authority to decide on the environmentally sustainable level of water that could be taken from the Basin. They also found that the hydrological models being used were among the best available, and that the methods of analysis and interpretation were sufficient to begin a scientifically-based adaptive management process. The CSIRO review did, however, identify gaps in (i) the scientific knowledge included, (ii) the potential of other possible modelling, and (iii) the limited use of expert scientific opinion in developing the *Draft Plan*. The review concluded that the allocation of water now specified in the *Draft Plan* would only meet a minority of the targets set by the authority in its original *Guide*.

The *Draft Plan* was put to public consultation between the end of November 2011 and early 2012, resulting in nearly 12,000 comments. Interested publics tended to congregate under existing banners, such as the New South Wales Irrigators Council, or formed new ones, such as the Basin Communities Association, both of which engaged in the debate at public meetings and through news media. A final *Plan* was then drawn up (MDBA, 2012), and on 22 November 2012 it was approved by the national parliament. The Bill for the *Plan* accepted the amended recommendation for a reallocation of 2750 GL/y, but added two important clauses: the first delayed any action for 7 years, and the second allowed the volume of water for environmental flows to be revised upwards or downwards at an appropriate future time for adaptive management. This provided a political compromise that moved action on the issue forward, albeit slowly, and provided the potential for 'learning by doing'.

In 2008, very soon after Australia's new national Labor Government made the move to nationalise the management of the Murray-Darling River Basin, it also launched a project to develop, for the first time, a national curriculum for all Australia's schools (Minister for Education, 2008).

The Australian National Science Curriculum and the Murray-Darling River Basin

The National Curriculum, a first for Australia, was to replace the diverse curricula that had hitherto been the province and responsibility of the six individual states and two territories. Science was among the first four subjects to be developed and was to be mandatory for all students during their first ten compulsory years of schooling. In the final 2 years of school, students could then choose further science studies in any of Biology, Chemistry, Earth Science and Physics.

The details of the Australian National Curriculum for Science were endorsed in 2011 (ACARA, 2014a) and it is now currently being implemented by the different state and territory authorities. This very new science curriculum provides a pertinent opportunity to analyse its educational intentions and intended learnings, and to appraise the extent to which, in the compulsory years, it could contribute to students' understanding of the Murray-Darling River Basin issue—and hence other complex SSIs. The specialised science for Years 11 and 12 are not included in the analysis below, since they are all optional studies, and none of them are chosen by a majority of Australia's students.

The Science Curriculum begins with a *Rationale* for the place of Science in the totality of school learning, and this is followed by year-by-year *Content Descriptions* of the intended science learnings. From the beginning of developing the new science curriculum it was decided the intended learning in this new science curriculum would be in three strands, *Science Understanding, Science Inquiry Skills* and *Science as a Human Endeavour*. The second and third strands were influenced by recent research interest in the Nature of Science and the promotion by the OECD's PISA project of 'Knowledge about science' alongside the more familiar 'Knowledge of science' (OECD, 2007). Having two strands that relate to the 'Knowledge about science' was innovative, and signalled the inclusion of more aspects of the Nature of Science and its social bases than have hitherto usually been included.

Early in development of the curriculum, a challenging statement for the *Rationale* was unanimously adopted that affirms that the learning of Science should be challenging and oriented outwards to applications in the wider society: Science is about "interesting and important questions", and should be related to "local, national or global issues".

A very substantial debate then took place among the team of advisers and developers about how the content for learning in Science should be conceived and listed (Fensham, 2013). This debate reflected the alternative views of scientific literacy (SL) that Roberts (2007) described as Vision I (learning content drawn from the disciplinary sciences) and Vision II (learning content drawn from relevant real world science and technology contexts). It was, indeed, a rerun of debates that have occurred in many countries about the curriculum for school science since the 1990s. The final decision was to list the science content in the key *Science Understanding* strand from a Vision I perspective. This decision reflected Bernstein's (1971) more general conclusion about curricula that the power of a few (in this case, the bureaucrats) will win over others (in this case, the scientific and science teaching experts) in defining what is valued knowledge. Furthermore, it was decided that the *Inquiry Skills* and *Human Endeavour* strands would not be tightly linked to the content knowledge in the *Science Understanding* strand.

Each of these features of the Science curriculum is now examined against the science that underpins the reference targets listed above for the Murray-Darling Basin issue.

The Science Curriculum and the Murray-Darling River Basin Targets

The Rationale

The *Rationale* makes a clear statement concerning the role of science education in decision making and that the Science curriculum aims to prepare students to make informed judgements about real world issues:

> Science provides a way of answering interesting and important questions about the biological, physical and technological world. The knowledge it produces has proved to be a reliable basis for action in our personal, social and economic lives; Science is a dynamic, collaborative and creative human endeavour arising from our desire to make sense of our world; The curriculum supports students to develop the scientific knowledge, understanding and skills to make informed decisions about local, national and global issues. (ACARA, 2014a)

This statement requires the teaching of science to equip students to make decisions with respect to SSIs of all types, including complex ones such as the Murray-Darling River Basin issue.

The Science Understanding Strand

Despite the outward-looking *Rationale*, the science content for learning in the *Science Understanding* strand is oriented inwards to the science disciplines and not outwards towards real world issues involving science and technology. The strand is therefore traditional rather than innovative. For each grade level (1–10) the content descriptions are listed under the four familiar disciplinary sub-headings of *Biological sciences, Chemicals sciences, Earth and Space sciences* and *Physical sciences*. Over the 10 year levels, the strand lists 58 separate significant science knowledge topics. However, no "interesting and important questions" involving this science content knowledge are asked, and no "local, national or global issues" for making decisions are identified. Table 1 lists the science content knowledge in the *Science Understanding* strand, by year level, that has some direct or indirect relationship to the environmental targets for the Murray-Darling river system's health.

Although at least one of these knowledge topics could be related to each of the nine environmental targets, they do not represent enough of the underpinning science to make these targets understandable. Furthermore, much of this science is intended to be learnt in the primary years, when such a complex national S&T issue is less relevant to the majority of Australian students.

Table 1 Science content knowledge (by year level) in the *Science Understanding* strand that has potential relevance to the environmental targets for the Murray-Darling River Basin

Year level	Content descriptions
Biological sciences	
1	Living things live in different places where their needs are met
4	Living things have life cycles
4	Living things, including plants and animals, depend on each other and the environment to survive
5	Living things have structural features that help them survive in their environment
6	The growth and survival of living things are affected by the physical conditions of the environment
7	Interactions between organisms can be described in terms of food chains and food webs and humans can affect these interactions
9	Ecosystems consist of communities of interdependent organisms and abiotic components of the environment; matter and energy flow through these systems
Earth sciences	
2	Earth's resources including water are used in a variety of ways
6	Sudden geological changes or extreme weather conditions can affect Earth's surface
7	Water is an important resource that cycles through the environment
10	Global systems, including the carbon cycle, rely on interactions involving the biosphere, lithosphere, hydrosphere, and the atmosphere
Physical sciences	
6	Energy from a variety of sources can be used to generate electricity

Comment

The Vision 1 framing of the *Science Understanding* strand in terms of separate science disciplines makes it likely that the majority of Australia's students would complete their school science without exposure to many key interdisciplinary scientific phenomena that underpin the Murray-Darling River Basin issue. Soil salinity, the cycling effects of drought and flood, the impact of hydroelectric generation on river floodplains and their forests, the evaporative loss from slow and fast moving rivers, and the movement of river materials, are some topics that are likely to fall through the gaps that result from this Vision I framing. In contrast, a Vision II framing would have been very likely to have Australia's river issues as a theme at some point in Years 7–10. This would have required science teachers to teach the related disciplinary and inter-disciplinary sciences that are integral to the health of Australia's river systems and to signal their socio-scientific consequences. Although the science of any of these systems can be as complex as the Murray-Darling River Basin, numerous sub-issues involving simpler amounts of science knowledge could be identified, which may be common across river systems.

In Bernstein's (1971) view of the curriculum as a public statement, the Vision I framing of the *Science Understanding* strand creates a science curriculum that is most useful to future disciplinary scientists. A Vision II framing of the science knowledge as practical knowledge for real world contexts (Layton, 1991) would have more directly addressed the needs of all students and been in line with the outward intention of the *Rationale*.

It is noteworthy that the new Geography Curriculum does have a Vision II framing. Its themes and topics cover both the physical and social environment (ACARA, 2014b). For Years 7–10, under the headings *Water in the world, Place and liveability, Biomes and food security* and *Environmental change and management,* there are 18 topics that could be directly related to the Murray-Darling River Basin. The *Rationale* for Geography also refers to the importance of links being made with Science, but this cross boundary linkage has not been reciprocated in the Science curriculum.

The Science Inquiry Strand

The *Science Inquiry* strand identifies its intended science skills under the headings of *Questioning and predicting, Planning and conducting, Processing and analysing data and information, Evaluating,* and *Communicating.* The framing of this set of learnings suggests an active practice of these skills in relation to authentic contexts. This is in line with Allchin's (2011) 'whole science' approach to the learning of these aspects of the *Nature of Science,* rather than the rule and rubric manner that has often been used to teach about science inquiry.

The extensive scientific data in the *Guide to the Proposed Murray-Darling Basin Plan* (MDBA, 2010) would provide students with opportunity to practise the skills listed under *Processing and analysing.* How the *Guide's* conclusions are drawn from these data would illustrate some skills under *Evaluating.*

The skills under *Communicating* are an innovation in this Science curriculum and the developmental intention for them is evident across the year levels:

- Years 5/6—Communicating ideas, explanations and processes in a variety of ways including multi-modal texts.
- Years 7/8—Communicating ideas, findings and solutions to problems using scientific language and representations using digital technologies as appropriate.
- Years 9/10—Communicating scientific ideas and information for a particular purpose including constructing evidence-based arguments and using appropriate scientific language, conventions and representations.

Teaching for each of these involves students practising to present their own ideas, but also analysing science communications from a variety of sources. The *Guide* (MDBA, 2010) and the *Draft Plan* (MDBA, 2011a, 2011b) provide both good and bad examples of science communicating. Students in Years 5/6 would appreciate the variety of ways a river system can be presented. By Years 9/10 students would benefit from comparing the evidence-based arguments for the science claims in the *Guide* with the apparent lack of explicitly scientific evidence in the *Draft Plan*.

In relation to "Communicating for a particular purpose", the different ways the *Guide's* recommendations are framed compared with the framing of the *Draft Plan*, again would provide a good example of the importance of framing a message to meet "a particular purpose", for example, a particular target audience. Ogawa's (2013) discussion of 'drivers' and 'targets' in communication of science in private and public domains is relevant to the practice of this skill.

Comment

The *Science Inquiry* strand of the new Australian curriculum, with its expanded sense of the Nature of Science does have considerable potential for the teaching and learning of a number of intended skills. These skills apply not only to the Murray-Darling River Basin issue, but also to many other complex SSIs. However, scientific modelling, a Nature of Science skill that is gaining prominence in the management of SSIs, is notably absent in this strand. Fortunately, the research literature does have suggestions as to how this omission might be remedied in subsequent revisions of the curriculum. For example, in a review of the use of models in school science, Gilbert, Boulter, and Rutherford (1998) found that the focus is almost always on 'model' as a noun, rather than as a predictive verb. It is in this predictive sense that 'modelling' occurs in the science of complex SSIs. Justi and Gilbert (2002), in a subsequent study, found that science teachers did recognise predictive modelling as a scientific skill, and it is on this ground that a strong case has since been made to include this skill in science education (Clement, 2008; Gilbert, 2004). Justi and Gilbert's two stage process for developing modelling in science education has been further developed (Fensham, 2014) to provide a fairly simple procedure for its teaching.

The Science As a Human Endeavour Strand

The content descriptions of science knowledge in the *Science as a Human Endeavour* strand are more developmentally described than those in *Science Understanding and are organised under two substrands—Nature and Development of Science and Use and Influence of Science*. Accordingly, only those for Years 7–10 that directly or indirectly relate to the environmental targets are listed in Table 2.

All ten of these intended learnings lend themselves to be practised through the context of a complex SSI, as they certainly do for the example of the Murray-Darling River Basin. The first three learnings under *Nature and development of science* are pertinent to the Basin's science, and integral to the allocation of water in the *Guide*. Their adequacy is also the subject of comment in the CSIRO review of the *Draft Plan*.

Table 2 Intended science learnings (by Year Level) in the *Science as Human Endeavour* (SHE) strand of direct relevance to the Murray-Darling issue

Year level	Content descriptions
Nature and development of science	
7 and 8	Scientific knowledge changes as new evidence becomes available, and some scientific discoveries have significantly changed people's understanding of the world
7 and 8	Science knowledge can develop through collaboration and connecting ideas across the disciplines of science
9 and 10	Scientific understanding, including models and theories, are contestable and are refined over time through a process of review by the scientific community
9 and 10	Advances in scientific understanding often rely on developments in technology and technological advances are often linked to scientific discoveries
Use and influence of science	
7 and 8	Science and technology contribute to finding solutions to a range of contemporary issues; these solutions may impact on other areas of society and involve ethical considerations
7 and 8	Science understanding influences the development of practices in areas of human activity such as industry, agriculture and marine and terrestrial resource management
7 and 8	People use understanding and skills from across the disciplines of science in their occupations
9 and 10	People can use scientific knowledge to evaluate whether they should accept claims, explanations or predictions
9 and 10	Advances in science and emerging sciences and technologies can significantly affect people's lives, including generating new career opportunities
9 and 10	The values and needs of contemporary society can influence the focus of scientific research

These learnings also acknowledge that the relevant science for an SSI can often be uncertain or incomplete, a feature of science that Kirch (2012) made a strong case for including as a learning goal for science education. She pointed out that the National Research Council in the USA, more than a decade earlier, had referred to uncertainties in science in the *Standards for Science Education* (National Research Council, 1996), and set out a two-pronged model for their teaching and learning based on their empirical aspects and their more psychological origins. Incidentally, the *Standards* were published in the same year that the international scientific community issued the *Precautionary Principle* (COMEST, 2005) as an approach to dealing with uncertainty in science for decision-making.

The Murray-Darling River Basin also offers examples that could apply to *Use and influence of science*. For example, that "Science and technology contribute to finding solutions to a range of contemporary issues"; that "These solutions may impact on other areas of society and involve ethical considerations"; and an example of "Science influencing agricultural practices" and "People using science in their occupations". "The use of science knowledge to evaluate claims" is clearly evident in the *Guide* but is underplayed in the *Draft Plan*. The creation of the Murray-Darling Basin Authority and its charter of responsibility illustrates the importance of "The influence that social values and needs have on the locus of scientific research".

It is the complexity of the Murray-Darling River Basin as an SSI that makes it so applicable to these learnings. Furthermore, the rich potential in the *Science as a Human Endeavour* strand is not restricted to the Murray-Darling River Basin issue. All ten of its learnings would be relevant in most other complex SSIs.

Comment

As we examine below, the case of the Murray-Darling River Basin exemplifies the range of scientific knowledge in a complex SSI and how its issues can become framed in different ways. Twenty-five years ago, Hardwig (1991) pointed out that many contemporary questions in science require the work of a number of different scientists who contribute their bit to the overall endeavour. It is common that none of them will be fully conversant with what the others are doing, but they must develop mechanisms of 'trust' within the science community. This statement also applies to the science underpinning complex SSIs, which invariably involves contributions from scientists from different disciplines, whose findings may have different levels of trust in the scientific community. Nevertheless, as in the Murray-Darling River Basin case, despite the variability of certainty and trust in the underlying science, political decisions may still be necessary.

Norris (1995) extended Hardwig's idea of 'trust' among scientists to science education by suggesting that 'intellectual dependence' is now a more realistic goal for school science than the 'intellectual independence' towards which he claimed school science has traditionally aimed. His radical ideas received little attention at the time, but now these ideas need revisiting in relation to the teaching of SSIs in science education. They do, however, need further explication, as they can be easily misunderstood. The intellectual independence, which he claimed has pertained to school science, is related to the sense of students in schooling being regarded as individuals, each expected to learn the science knowledge on offer. The constructivist pedagogies for teaching and learning that were popularised among science educators in the 1980s also treated the student as an individual developing knowledge. The later notions of social constructivism acknowledged students as more of learning a community.

In another sense, traditional science education has made students dependent on 'Science' as the bounded and established knowledge someone else has decided to include in the curriculum. They have had no independence about what or how to learn in school science. In whichever of these senses we describe the learning of the introductory pieces of disciplinary science in a typical science curriculum, Norris is arguing that a new term is needed for students' relation to the very diverse and often uncertain science in real world S&T contexts (and SSIs). He suggests that a stance of active intellectual dependence is a realistic one for science teachers and their students to adopt. This stance is not at all a passive one, as the word 'dependence' may imply, but is very active because it requires science students to learn who can, and cannot, be trusted regarding scientific claims to knowledge, and to know how to make judgements about credibility. This expertise, Norris foreshadows, will involve (i) learning science content, (ii) learning about science (its philosophical basis, historical progression, and social processes), and (iii) learning to live with science as an important—but not the only—source of knowledge. Each of these loomed large in the case of the Murray-Darling management issue.

The decision of the Australian curriculum authorities to give separate status to the three curriculum strands is fortuitous since it does not tie the second and third strands to the familiar *Science Understanding* that, as shown above, is least relatable to SSI teaching. Australian science teachers are, thus, free to choose both intra-science and out-of-school S&T contexts (e.g., SSIs) for teaching the range of skills and learnings that these two more helpful strands (*Science Inquiry Skills* and *Science as a Human Endeavour*) intend.

The success of this set of largely new science learnings will, however, very much depend on the support Australian science teachers receive in professional development and from the assessment authorities. Since these rather novel skills have not been part of the usual programmes for professional development, there will be a particular need for authorities to harness the help of science education researchers who have become aware of exemplary pedagogies that have been found in research studies to develop these skills. Similarly, the authorities responsible for assessing learning will need to publicise and use authentic modes for assessing the learning of these Nature of Science skills in context. A range of alternative modes for these sorts of assessment are also now appearing in in the literature (Allchin, 2011; Fensham & Rennie, 2013; Sadler & Zeidler, 2009). Since 2000, the OECD's PISA Science project, despite being restricted to paper and pencil testing, has also used contextually–based items to measure the learning of inquiry-based and evidence-based scientific literacies (OECD, 2007).

We now turn to the analysis of how the *Draft Plan* for the management of the Murray-Darling River Basin was reported and debated in the public domain. The role and importance of framing in this communication provides an exemplary means of access for bringing the complexity of this and other SSIs into the classroom.

Public Deliberation and the Murray-Darling River Basin

In the public domain the purely biophysical basis of an SSI loses much of its science disciplinary boundaries. The range of interest groups that are mobilised around issues such as the management of the Murray-Darling River Basin have diverse perspectives on what counts as evidence (scientific and other) and which experts are considered relevant. In order to examine the breadth of some of these perspectives around the issue of water management in the Murray-Darling, we analysed both published official documents and national newspaper accounts. Although not comprehensive in coverage, these statements provide an indication of how different communities framed the issue and conducted boundary work around evidence and expertise in the process. Forty-three newspaper articles or letters were sourced from the online repository of one of Australia's largest circulation newspapers, the Sydney Morning Herald, between the release of the *Draft Plan* (28 November 2011) and its final approval by national parliament (22 November 2012).

The analysis involved the extraction of quotes for each position-statement in the documents, articles and letters and assigning them to a category. Other perspectives did, of course, exist and they were often expressed in other sources like talkback radio. Our approach was not comprehensive, nor does it provide a measure of the dominance or relative weight of each of the categories. More simply, it reveals something of the range of perspectives that were expressed in the debate. In a similar way, any complex SSI will generate diverse responses in the public arena.

Framing of the Water Allocation Figure Among Different Groups

Throughout the debate and public consultation on the *Draft Plan*, the water allocation figure of 2750 GL/y became a central focus of division. However, the disagreement about this figure was not a straightforward disagreement about its underlying scientific basis. Rather, it was evident that an apparently scientific value such as the amount of water needed for healthy environmental flow can be understood as both 'sound science' and as 'political compromise'—a dualism that has been found in other complex SSIs.

The initial framing of the Murray-Darling River Basin as a regulatory issue goes back to the earlier attempt in 2003 at management of the river system. Crase et al. (2005) and Crase, O'Keefe, and Dollery (2013), in their study of the public debate that occurred then, pointed out that its focus on a fixed allocation of water allowed critics to claim that other important attributes of the issue had not been taken into account. Rather than simply asking a regulatory question (i.e., What is sufficient environmental flow to comply with the Water Act?), different groups queried what allocation of water was politically feasible, economically sensible and culturally appropriate.

When the question is rephrased in this way, a regulatory framing of the issue is no longer sufficient and a much broader range of evidence is now needed. Nevertheless, in the public debate following the release of the *Guide* in 2010, the Basin Authority continued to frame the issue in regulatory terms by releasing just the scientific reasoning behind the *Guide's* proposed water allocations.

This regulatory framing of the issue drew positive responses from scientific groups and environmentalists, but very negative responses from farming communities and other stakeholders, who argued it overlooked their needs and interests (Wroe, 2011). In their updated 2011 *Draft Plan*, the re-constituted Authority still presented a water allocation figure, but this time not only justified it on scientific grounds, but also on social and economic modelling (MDBA, 2011a, 2011b). The recommended reduction in the amount of water to be reallocated to environmental flows was an explicit compromise, intended to be part of on-going adaptive management of the river. However, this multiple framing of the issue as both a regulatory necessity and a socioeconomic compromise drove a persistent rift between the different interest groups.

On one hand, many scientists and environmentalists who had originally supported the water allocation in the 2010 *Guide* and the science that underpinned it now attacked the revised water allocation. One of the major groups was the Wentworth Group of Concerned Scientists, a self-assembled group of scientists, economists and business people who had interest and expertise in the management of Australia's natural resources. Although not exclusively a scientific body, the Wentworth Group was founded on the basis that they would "connect science to public policy" (wentworthgroup.org). The prominence of its scientific members in other science bodies lent the Group clout as a worthy voice in the debate. In their response to the *Draft Plan*, the Wentworth Group argued that the *Plan* lacked sufficient scientific information, made unjustified assumptions about the sustainability of ground water, and neglected the impact of climate change. In short, the revised allocation of water lacked what they considered a credible scientific base of evidence:

> The science used to establish the evidence for the 2,750 GL reduction is not only absent from the documentation, but even more disgraceful is that the science for the 2,750 GL reduction is not accorded the scientific scrutiny of transparent independent review. It is impossible to assess the ecological outcomes from a reduction to extractions of 2,750 GL from the information in these tables... Without the information to assess this, it is impossible to determine whether the draft Basin Plan complies with the Water Act. (Cosier et al., 2012, p. 10)

This perspective maintained and reinforced the regulatory frame of the debate that had been dominant in the original *Guide*.

In contrast, local community groups and irrigators showed concern about the economic and social impacts of the *Draft Plan*. Rather than challenge the scientific basis for and the limited environmental impact of the 2750 GL/y figure, their socioeconomic frame suggested that there still had been insufficient cost-benefit analysis to justify this amount of reallocation. Rather than questioning the validity of the science on its biophysical basis, this group argued that other important attributes and impacts had still not been adequately included. For example:

> The concern we've got primarily is that [the MDBA] haven't looked at people, profit and the planet... What they've done is they've looked at a cost-benefit of the environment and ignored people and profit. (NSW Farmers' Association chief executive, quoted in SMH, 2011a)

Comment

According to these media reports, the fixed water allocation figure could be challenged in different ways under both a regulatory and socioeconomic frame. One focused on the biophysical bases of a healthy river, and the other focused on the socioeconomic impacts on communities and livelihoods. Despite a shared disapproval of the proposed volume of water, the two sides drew on different forms of evidence and expertise to support their case. Their arguments, therefore, became mutually incompatible.

The challenges associated with a lack of mutual understanding have been previously explored by Lock (2011), who suggests that scientists and other publics should clearly communicate the evidentiary bases of their positions. Only then can 'talking past each other' be avoided and productive dialogue be achieved. In the case of the *Draft Plan*, the Authority's maintenance of focus on the water return of 2750 GL, albeit now using a regulatory frame, meant this miss-communication was indeed the case.

Drawing Boundaries Around Scientific Evidence and Authority

The theoretical approach of 'boundary work', set out earlier, provides a tool for looking at the way in which lines are drawn around science and scientists in society, and why this distinction becomes important in making sense of SSIs.

In the public deliberations about an SSI, there is often a series of competing claims to scientific authority. Who, then, can be regarded as a scientific expert? For example, in its highly critical response to the Basin Authority's *Draft Plan*, the Wentworth Group challenged its scientific authority by arguing that the Authority "manipulates science" for a "pre-determined political outcome":

> The Murray-Darling Basin Authority ignores much of the good work and has instead produced a draft Plan that manipulates science in an attempt to engineer a pre-determined political outcome. The Commonwealth government should stop the process, instruct the Authority to withdraw the draft Plan, abandon the proposal for a 2015 review and instead take the time necessary to include the science and social science now. (Cosier et al., 2012, p. 1)

Using their claimed status as 'concerned scientists', the Wentworth Group challenged the scientific legitimacy of the Basin Authority, excluding them from the boundary of science and portraying them as politically motivated. However, this charge was soon counteracted by the Authority's chairman who employed his own boundary work to undermine the claimed scientific authority of the Wentworth Group:

> The views of the Wentworth Group are well known. As with other groups with diametrically opposed opinions on the Draft, all views will be considered as part of the consultation period. (Craig Knowles, MDBA chairman, quoted in Arup, 2012)

Interestingly, few other explicitly scientific voices were given coverage in the newspaper articles on the *Draft Plan*. This may be because individual scientists were reluctant to enter public debate, or because it was perceived that the Wentworth Group already represented a 'universal' scientific position. In other science-related controversies, the absence of scientific voices in the public arena has had the consequence that non-scientific voices are able to advocate on behalf of science (Gregory & Lock, 2008; White, 2011).

In the case of the Murray-Darling River Basin, the absence of scientific voices meant that judgement of the Authority's *Draft Plan* was left predominantly to non-scientists. One article reported that irrigators gave the Draft Plan "a 'fail' rating on six out of seven criteria such as transparency, detail and balance" (Wroe, 2011). Another reported on a politician from the Australian Greens Party who argued that the plan "will fail to save the river and the species that rely on it" (SMH, 2011b). As in the earlier Murray-Darling debates (Crase et al. 2005), the absence of government socioeconomic data sources enabled some lobby groups to produce and publicise their own figures without independent verification. The validation of evidence was no longer the province of the scientific community, but had moved to other social actors for judgement and debate.

Conclusion

The management of the Murray-Darling River Basin illustrates the complexity of many SSIs. These issues can become seen from multiple perspectives, and public deliberation can demand evidence that extends far beyond a purely scientific or technical basis. At the same time, SSIs mobilise a range of stakeholders each with their own claim to expert authority on adjudicating how a controversy might find closure.

In the science classroom attention should be drawn to the diverse aspects of SSIs, including those that emphasise the non-science aspects. In doing so, students could be encouraged to view the issue from the point of view of different stakeholders and, in a role playing sense, students could be assisted to present the issue from the different perspectives that emphasise its social, economic, environmental or moral aspects, including what evidence and expertise might be relevant in each case.

With respect to the scientific bases of complex SSIs, the new Australian National Science Curriculum offers considerable opportunity for science teachers to include complex SSIs among the contexts they explore in their science education. There are, however, some aspects of SSI science that stand out as challenges that are yet to gain authoritative approval in the science curriculum.

Opportunities and Challenges

Despite the clearly stated intention in the *Rationale* of the Australian Science Curriculum for engagement with SSI issues, no such issues are suggested as examples. Instead, the manner in which the detailed knowledge for learning is listed, at best, allows science teachers to choose one piece of this knowledge as a starting point to open their students to complex SSIs and, at worst, <u>discourages</u> them from doing so. The *Science Understanding* strand is especially deficient. Its disciplinary listing of science content—a Vision I framing—fails to recognise much relevant interdisciplinary science. A science curriculum that aims to equip students, as future adult citizens, to understand and make decisions about complex SSIs should point to exemplary SSIs and list some of their disciplinary and interdisciplinary science content. A more thematically designed curriculum—one with a Vision II framing— would <u>encourage</u> and <u>require</u> science teachers to use some of their classroom time engaging with these issues.

The *Science as Inquiry* and *Science as Human Endeavour* strands of the Australian Science Curriculum, by extending students' understanding of the Nature of Science, do offer considerable opportunity for teachers and students to practice scientific skills and intellectual procedures that are integral to complex SSIs. The strands do not, however, include two key scientific aspects of complex SSI, namely, the skill of modelling and issues concerning of the certainty/uncertainty of scientific knowledge and its warrants for trust.

A big challenge associated with the opportunity to teach these skills and processes of the Nature of Science is a pedagogical one. Their teaching and learning will require science teachers to use dialogical pedagogies in their classrooms, which are very different from the transmissive ones so often used when science content knowledge alone is the central focus. For example, the new emphasis on science communication as a skill will require students to practice alternative ways of framing the same science for different purposes and audiences. As highlighted by Gregory and Lock (2008), public engagement with science is not just about the public developing an understanding of the science, but it is also an opportunity for scientists to "listen and learn as well as speak and teach" (p. 1257, see also Pedretti & Navas-Iannini, chapter "Pregnant Pauses: Science Museums, Schools and a Controversial Exhibition" and Stocklmayer, chapter "Communicating Science", this volume). This dictum applies also to teachers in science classrooms.

Finally, if the authorities responsible for the school science curriculum are serious about the curriculum's stated intention to bring "decision making about SSIs" into the classroom, they will need to respond to three obvious challenges:

- to recast the curriculum so that this intention is given priority,
- to develop new means for the assessment of science learning to ensure this priority is reinforced, and
- to ensure that science teachers get the support in professional development support they will need for these new teaching tasks.

Each of these will involve a considerable amount of revisionary boundary work in relation to science education. Only then can science teachers be expected to likewise change their sense of the boundary of science and engage with their students in making decisions about these far-reaching socio-scientific issues.

References

AAAS. (2006). *Annual report*. Washington, DC: Author.

ACARA. (2014a). *The Australian curriculum: Learning areas: Science*. Retrieved from http://acara.edu.au/curriculum_1/learning_areas/science.html

ACARA. (2014b). *The Australian curriculum: Learning areas: Humanities and social sciences: Geography*. Retrieved from http://acara.edu.au/curriculum_1/learning_areas/humanitiesand-socialscience/geography.html

Aikenhead, G. S. (1992). Logical reasoning in science and technology. *Bulletin of Science, Technology and Society, 12*, 149–159.

Allchin, D. (2011). Evaluating knowledge of the nature of (whole) science. *Science Education, 95*(3), 518–542.

Arup, T. (2012, January 19). Scientists reject plan to save Murray-Darling. *The Sydney Morning Herald*. Online.

Bell, R. L., & Lederman, N. G. (2003). Understandings of the nature of science and decision making on science and technology based issues. *Science Education, 87*, 352–377.

Bencze, L., & Carter, L. (2011). Globalising students acting for the common good. *Journal of Research in Science Teaching, 48*(6), 648–669.

Bernstein, B. (1971). On the classification and framing of educational knowledge. In M. F. D. Young (Ed.), *Knowledge and control* (pp. 47–69). London, UK: Collier-Macmillan.

Bonneuil, C., Joly, P.-B., & Marris, C. (2008). Disentrenching experiment: The construction of GM crop field trials as a social problem. *Science, Technology & Human Values, 33*, 201–229.

Clement, J. (2008). *Creative model construction in scientists: The role of analogy, imagery and mental stimulation*. Dordrecht, The Netherlands: Springer.

Collins, H. (2009, March 5). We cannot live by scepticism alone. *Nature, 458*, 30–31.

COMEST. (2005). *The precautionary principle*. Paris, France: UNESCO.

Cosier, P., Davis, R., Flannery, T., Harding, R., Hughes, L., Karoly, D., … Williams, J. (2012). *Statement on the 2011 draft Murray-Darling Basin plan*. Sydney, Australia: Wentworth Group of Concerned Scientists.

Crase, L., Dollery, B., & Wallis, J. (2005). Community consultation in public policy: The case of the Murray-Darling Basin of Australia. *Australian Journal of Political Science, 40*(2), 221–237.

Crase, L., O'Keefe, S., & Dollery, B. (2013). Talk is cheap, or is it? The cost of consulting about uncertain reallocation of water in the Murray–Darling Basin, Australia. *Ecological Economics, 88*, 206–213.

CSIRO. (2011). *The Murray-Darling Basin science*. Retrieved from www.csiro.au/science/MDBscience

Driver, R., Newton, P., & Osborne, J. (2000). Establishing the norms of scientific argumentation in classrooms. *Science Education, 84*, 287–312.

Engle, R. A. (2006). Framing interactions to foster generative learning: A situative explanation of transfer in a community of learners classroom. *Journal of the Learning Science, 15*(4), 451–498.

Fensham, P. J. (1985). Science for all: A reflective essay. *Journal of Curriculum Studies, 17*(4), 415–435.

Fensham, P. J. (2013). The science curriculum: The decline of expertise and the rise of bureaucratise. *Journal of Curriculum Studies, 45*(2), 152–168.

Fensham, P. J., & Rennie, L. J. (2013). Towards and authentically assessed science curriculum. In D. Corrigan, R. Gunstone, & A. Jones (Eds.), *Valuing assessment in science education: Pedagogy, curriculum, policy* (pp. 69–100). Dordrecht, The Netherlands: Springer.

Funtowicz, S. O., & Ravetz, J. R. (1993). Science for the post-normal age. *Futures, 25*(7), 739–755.

Gieryn, T. F. (1983). Boundary-work and the demarcation of science from non-science: Strains and interests in professional ideologies of scientists. *American Sociological Review, 48*(6), 781–795.

Gilbert, J. K. (2004). Models and modelling: Routes to more authentic science education. *International Journal of Science and Mathematics Education, 2*(2), 115–130.

Gilbert, J. K., Boulter, C., & Rutherford, M. (1998). Models in explanations: Horses for courses? *International Journal of Science Education, 20*(1), 83–97.

Gilbert, J. K., & Stocklmayer, S. (Eds.). (2013). *Communication and engagement with science and technology: Issues and dilemmas: A reader in science communication.* New York, NY: Routledge.

Goffman, E. (1974). *Frame analysis: An essay on the organization of experience.* Cambridge, MA: Harvard University.

Gregory, J., & Lock, S. J. (2008). The evolution of 'public understanding of science': Public engagement as a tool of science policy in the UK. *Sociology Compass, 2*(4), 1252–1265.

Hardwig, J. (1991). The role of trust in knowledge. *The Journal of Philosophy, 88,* 693–708.

Hodson, D. (2003). Time for action: Science education for an alternative future. *International Journal of Science Education, 25*(6), 645–670.

Hodson, D., Bencze, L., Elshof, L., Pedretti, E., & Nyhof-Young, J. (Eds.). (2002). *Changing science education through action research: Some experiences from the field.* Toronto, Canada: University of Toronto.

Irwin, A., & Wynne, B. (1996). *Misunderstanding science? The public reconstruction of science and technology.* Cambridge, UK: Cambridge University.

Jasanoff, S. (1987). Contested boundaries in policy-relevant science. *Social Studies of Science, 17,* 195–230.

Justi, R., & Gilbert, J. K. (2002). Modelling teachers' views on the nature of modelling and implications for the education of modellers. *International Journal of Science Education, 24*(4), 369–387.

Kirch, S. (2012). Understanding scientific uncertainty as a teaching and learning goal. In B. J. Fraser, K. Tobin, & C. McRobbie (Eds.), *Second handbook of research in science education* (pp. 851–864). Dordrecht, The Netherlands: Springer.

Kolstø, S. D. (2001). "To trust or not to trust....": Pupils' ways of judging information encountered in a socio-scientific issue. *International Journal of Science Education, 23*(9), 877–902.

Kortland, K. (1996). An STS case study about students' decision making on the waste issue. *Science Education, 80,* 673–689.

Layton, D. (1991). Science education and praxis: The relationship of school science to practical action. *Studies in Science Education, 19,* 43–79.

Layton, D., Jenkins, E., Macgill, S., & Davey, A. (1993). *Inarticulate science? Perspectives on the public understanding of science and some implications for school science.* Driffield, UK: Studies in Education.

Levinson, R. (2004). Teaching bioethics in science: Crossing a bridge too far? *Canadian Journal of Science, Technology and Mathematics Education, 4,* 353–369.

Levinson, R. (2006). Towards a theoretical framework for teaching controversial socio-scientific issues. *International Journal of Science Education, 28*(10), 1201–1224.

Levinson, R. (2010). Science education and democratic participation: An uneasy congruence. *Studies in Science Education, 46*(1), 69–119.

Lock, S. J. (2011). Deficits and dialogues: Science communication and the public in the understanding of science in the UK. In D. J. Bennett & R. C. Jennings (Eds.), *Successful science communication: Telling it like it is* (pp. 17–30). Cambridge, UK: Cambridge University.

MDBA. (2010). *Guide to the proposed basin plan: Overview.* Canberra, Australia: Author.

MDBA. (2011a). *Plain English summary of the proposed basin plan—Including explanatory notes.* Canberra, Australia: Author.

MDBA. (2011b). *Socioeconomic analysis and the draft plan: Part A—Overview and analysis.* Canberra, Australia: Author.

MDBA. (2012). *Proposed basin plan consultation report.* Canberra, Australia: Author.

Minister for Education. (2008, October 12). Media release. *Delivering Australia' first national curriculum.*

National Research Council. (1996). *National science education standards*. Washington, DC: National Academics.

National Research Council. (2001). *Grand challenges in environmental sciences*. Washington, DC: Author.

Nelkin, D. (1979). *Controversy: Politics of technical decisions*. Beverly Hills, CA: Sage.

Norris, S. (1995). Living with scientific expertise: Towards a theory of intellectual communalism for guiding science teaching. *Science Education, 79*(2), 201–217.

OECD. (2007). *PISA 2006 science competencies for tomorrow's world. Vol.1. Analysis*. Paris, France: OECD.

Ogawa, M. (2013). Towards a 'design approach' to science education. In J. K. Gilbertt & S. Stocklmayer (Eds.), *Communication and engagement with science and technology: Issues and dilemmas: A reader in science communication* (pp. 3–18). New York, NY: Routledge.

Patchen, T., & Smithenry, D. W. (2013). Framing science in a new context: What students take away from a student-directed inquiry curriculum. *Science Education, 97*(6), 801–829.

Poff, N. L., Allan, J. D., Palmer, M. A., Hart, D. D., Richter, B. D., Arthington, A. H., et al. (2003). River flows and water wars: Emerging science for environmental decision making. *Frontiers in Ecology and the Environment, 1*(6), 298–306.

Ratcliffe, M. (1997). Pupil decision making about socio-scientific issues within the curriculum. *International Journal of Science Education, 19*(2), 167–182.

Rein, M., & Schön, D. (1994). *Frame reflection: Towards the resolution of intractable policy controversies*. New York, NY: Basic Books.

Rittel, H., & Webber, M. (1973). Dilemmas in a general theory of planning. *Policy Sciences, 4*, 155–169.

Roberts, D. (2007). Scientific literacy/scientific literacy. In S. K. Abell & N. G. Lederman (Eds.), *Handbook of research on science education* (pp. 125–177). Mahwah, NJ: Lawrence Erlbaum.

Ryder, J. (2003). Identifying science understanding for functional scientific literacy. *Studies in Science Education, 36*, 1–44.

Sadler, T. D., & Zeidler, D. L. (2009). Scientific literacy, PISA and socio-scientific discourse: Assessment for progressive aims of science education. *Journal of Research in Science Teaching, 46*(8), 909–921.

SMH. (2011a, November 28). Murray-Darling plan 'ignores' NSW farmers. *Sydney Morning Herald*. Online.

SMH. (2011b, November 28). Murray-Darling water plan mired in controversy. *Sydney Morning Herald*. Online.

Solomon, J., & Aikenhead, G. (1994). *STS education: International perspectives on reform*. New York, NY: Teachers College.

Sullivan, C. A. (2014). Planning for the Murray-Darling Basin: Lessons from transboundary basins around the world. *Stochastic Environmental Research and Risk Assessment, 28*, 123–136.

Thomson, S., Hillman, K., & Wernert, N. (2012). *Monitoring Australian year 8 students outcomes internationally*. Camberwell, Australia: ACER.

White, S. (2011). Dealings with the media. In D. J. Bennett & R. C. Jennings (Eds.), *Successful science communication: Telling it like it is* (pp. 151–166). Cambridge, UK: Cambridge University.

Wroe, D. (2011, December 5). Murray-Darling proposal slammed by irrigators. *Sydney Morning Herald*. Online.

Zeidler, D. L., & Sadler, T. D. (2008). The role of moral reasoning in argumentation: Conscience, character, and care. In S. Erduran & M. P. Jiménez-Aleixandre (Eds.), *Argumentation in science education: Recent developments and future directions* (pp. 201–216). New York, NY: Springer.

Zeidler, D. L., Sadler, T. D., Simmons, M. L., & Howes, E. V. (2005). Beyond STS: A research-based framework for socioscientific issues education. *Science Education, 89*(3), 357–377.

Outreach Education: Enhancing the Possibilities for Every Student to Learn Science

Debra Panizzon, Greg Lancaster, and Deborah Corrigan

Abstract A review of the science education literature identifies the importance of outreach in raising public awareness of science while providing students with contextually relevant and meaningful science in ways that enhance their school experiences. The National Virtual School of Emerging Sciences (NVSES) provided just such an opportunity. Established throughout 2012–2014, it enabled 429 secondary students from across Australia to engage with the emerging sciences of Astrophysics and Nanotechnology. Creation of 'virtual' science classrooms allowed small groups of students to connect synchronously twice a week under the guidance of subject specialist teachers. To prepare for this context, teachers modified their face-to-face pedagogies to suit the range of technologies readily accessible in the virtual classroom. This chapter discusses how these different pedagogies were utilised by the NVSES teachers to develop lessons that created unique experiences for students within the virtual classroom environment. Data collected from pre and post student surveys, interviews with the NVSES teachers and access to digitally-recorded lessons demonstrate that while NVSES was highly successful, there were challenges for all involved.

Keywords Outreach education · Virtual learning environments · Learning science

Introduction

A review of the science education literature identifies the importance of various forms of outreach (as defined in chapter "Navigating the Changing Landscape of Formal and Informal Science Learning Opportunities") in raising public awareness of science while providing students with opportunities to experience contextually relevant and meaningful science in very different ways to formal school science

D. Panizzon (✉) · G. Lancaster · D. Corrigan
Monash University, Clayton, VIC, Australia
e-mail: debra.panizzon@monash.edu; greg.lancaster@monash.edu; deborah.corrigan@monash.edu

© Springer International Publishing AG, part of Springer Nature 2018
D. Corrigan et al. (eds.), *Navigating the Changing Landscape of Formal and Informal Science Learning Opportunities*,
https://doi.org/10.1007/978-3-319-89761-5_9

(Braund & Reiss, 2006; Falk & Dierking, 2000; Hodson, 1998). Historically, these science outreach experiences have often been viewed by the public as lying at one end of a spectrum with school science at the other, setting up a clear dichotomy. However, with increased access to these outreach providers via the Internet through virtual tours and online interactions while in the classroom, this dichotomy might be more productively conceived as a continuum of science learning opportunities (Malcolm, Hodkinson, & Colley, 2003).

An alternative way of thinking about these various outreach experiences is provided by Stocklmayer, Rennie, and Gilbert (2010) with their emphasis on the *learning* rather than the setting or place (e.g., school versus zoo). They support the view that there are many shared elements about learning in science regardless of the setting, including the need for clarity about the purpose and process; ensuring personal relevance; using science in local contexts; and some degree of choice by the learner. Rennie (2007) also identify clear distinctions about the contexts of informal settings that make them quite *unique*: (i) voluntary attendance and involvement; (ii) the curriculum (if evident) is open, offering choice to the learner; (iii) activities in which the learner is involved are not evaluated or assessed so they are non-competitive; and (iv) interaction is not homogenous being across age groups. So, a secondary teacher organising a fieldtrip to the local zoo with a clear agenda to link with what has been taught in school with a specific assessment task occurring at the end does not fit with this view of informal learning. This distinction, made by Stocklmayer et al. (2010), is critical in placing the focus on the *learning* rather than assuming that a change in context will merely result in learning. With this point considered, there is corroboration in the research regarding the potential benefits that outreach opportunities provide in supporting and enhancing students' understandings of science and scientific processes while generating interest and engagement with the general public.

In this chapter we discuss the key points around outreach in Australia that emerged from a recently published audit. Following this we introduce the National Virtual School of Emerging Sciences (NVSES) that was deliberately established to bridge the gap between traditional school science lessons and outreach by enabling students, regardless of their geographical location, the chance to connect with like-minded students in a virtual classroom using an online platform. In most cases, participation was voluntary, giving students access to experiences outside of the required curriculum and involved teachers and scientists who were experts in their fields. Initially, the context of NVSES is presented to provide a background to the students who participated in the programme. Subsequently, the role of specialist science educators as 'curators of resources' is explored along with the learning opportunities available to students through various digital technologies. Finally, some of the key challenges met by the team in designing and implementing NVSES with students and their participating schools are elaborated and discussed.

Outreach and the 'State of Play' in Australia

The level and diversity of outreach available in Australia was synthesised in a recent report entitled a *National Audit of Australian Engagement Activities 2012* (Metcalfe, Alford, & Shore, 2013). The report provides the first national picture of 411 science outreach or "engagement activities" (p. 1) that were available from January 2011 to June 2013 through various educational centres (e.g., museums), institutions (e.g., universities), and research organisations (e.g., Commonwealth Scientific and Industrial Research Organisation, CSIRO) as part of National Science Week. Data for the audit were collected from 254 respondents using an online survey that captured a range of information including details about: (i) the nature of the activity; and (ii) how the activity aimed to engage Australians with scientific issues and increase national/international interest in science.

Regarding the *Nature of the activity*, the top 10 types of activities in order of frequency from highest to lowest scoring were:

- Presentation/seminar/lecture
- Educational/school-based activity (e.g., participation in a progamme at school)
- Visit/tour, including school visits (e.g., Scitech Outreach Science workshops)
- Professional development workshop/course (e.g., Science Discovery Club)
- Hands-on activities (e.g., Science Fair)
- Show/demonstration
- Exhibit/poster (e.g., photography exhibition at a museum)
- Publication/tool (e.g., development of workbook and activity ideas on local plants and animals)
- Quiz/competition (e.g., Crest Awards)
- Art and science interaction (e.g., Concept Radical—an art competition that called for artists' impressions of free radicals).

As part of the survey, respondents had to specify how they engaged their audience in the activity. Metcalfe, Alford, & Shore, (2013) found that the majority of activities involved one-way communication with participants expected to "learn from watching, listening or viewing" (p. 14), with the methods cited including websites, newsletters, brochures, seminars and exhibitions. The practical activities (i.e., shows/demonstrations) were either science communicator or educator-directed, or required students to complete projects (e.g., the Crest Awards funded by the CSIRO). Importantly, though, these practical activities did encourage participants to "ask questions" and "share their views" with the opportunity to "problem-solve" while undertaking the project-based work (p. 14). Another interesting finding regarding the nature of the activities was that most were targeted at school-aged children and the science was focused on biological or environmental topics. While physics, chemistry, agriculture, and mathematics were addressed in activities, this was to a lesser degree with only a small proportion of activities focusing on engineering and information technology (including computing). Reflecting on this latter finding, we find it is surprising that information technology did not hold a more predominant focus within the outreach activities given its increasing impact on society.

In considering the overall findings, Meltcalfe et al. (2013) identified the following key points in their recommendations:

We recommend encouraging and supporting science activities that go for a longer time and which focus more on:

1. Specific groups rather than the general public or school children (e.g., farmers or youth aged 16–18)
2. The uptake rather than the delivery of science
3. Group problem solving
4. Consulting and sharing views about science
5. Activities that shape science questions
6. Critical thinking and dialogue about science
7. Achieving behavioural or policy change. (p. 54)

While the audit report is vague in the level of detail it provides for each activity and the data, it does provide at least a snapshot around the diversity of activities available at a point in time (i.e., National Science Week 2012) and some insight about what was likely experienced by the participants. With this said, the findings are somewhat disappointing given the emphases on what might be considered as fairly 'traditional' ways for engaging the public with science (see chapter "Communicating Science" by Sue Stocklmayer). However, the development of the NVSES model, outlined below, may signal the beginning of more innovative approaches to addressing some of these challenges.

Creating 'Virtual' Outreach Experiences for Students

The National Virtual School of Emerging Sciences (NVSES) was designed to enrich the learning opportunities for predominantly Year 10 students in the areas of Astrophysics, Quantum Physics, Nanoscience, and Nanotechnology. The 'virtual' classrooms were created using Cisco WebEx®[1] video conferencing software, allowing students in schools across all states and territories in Australia to connect with their peers through personal computers twice a week over a period of 8 weeks. The two synchronous lessons were team-taught by specialist teachers in the various fields with a teacher to student ratio of up to 1:25. As part of the experience, students were able to use their webcams and microphones to connect visually, listen using individual headsets, raise their symbolic hands (on their computer) so that teachers could ask individual students to respond, chat with other students in an open text-based forum termed 'back chat', and access shared documents through Google Drive using Google Docs (e.g., powerpoint presentations, documents and spreadsheets).

---

[1] Registered platform of Cisco International.

Table 1 Total enrolments in NVSES virtual classrooms

		Astrophysics	Quantum physics	Nanoscience	Nanotechnology
2013	Term 1	26	[a]NA	[a]NA	[a]NA
	Term 2	[a]NA	24	12	[a]NA
	Term 3	90	[a]NA	65	[a]NA
	Term 4	10	13	[a]NA	10
2014	Term 1	18	25	9	3
	Term 2	19	24	4	3
	Term 3	13	35	12	15

[a]*NA* not available for enrolment

Funded by the Federal Government, NVSES was designed and implemented by the authors in conjunction with a larger project team. We also carefully monitored, researched and evaluated the project. A range of data was collected including pre and post online questionnaires from students at the beginning and end of each teaching term (i.e., 8 weeks in length), an email survey completed by teachers in participating schools, and interviews with the specialist teachers conducting the NVSES lessons. These data are used throughout this chapter to provide evidence about the student experience.

NVSES was designed and implemented to enhance the experiences of students in areas of emerging sciences—not to replace school science. During the course of a 21-month period, a total of 329 students representing 46 schools across Australia participated in the virtual classrooms in the four units as they became available. However, over 100 of these students participated in more than one unit with a total enrolment of 430 over the period (Table 1). As can be observed from the enrolments, students demonstrated a preference for Astrophysics and Quantum Physics units over the Nanoscience and Nanotechnology. A survey of teachers in participating schools suggested that this might be partially due to the availability of these 'Nano' subjects in Year 10 chemistry (especially in Victorian schools) so these may not have been perceived by students as offering 'new' scientific opportunities.

As mentioned above, NVSES allowed students across Australia to connect to the synchronous lessons. The highest proportion of students (67%) enrolled from Victoria, which was the hosting state for NVSES. South Australia provided 14% of students, New South Wales 7%, Australian Capital Territory and Tasmania 5% each, with Queensland and Western Australia contributing 1% of students. While only a small proportion, participation by students from Western Australia was significant given that WA is 2 h behind Victoria with Australian Eastern Standard Time. The significance of this will become apparent with further information regarding how NVSES was conducted. Another important aspect was that 62% of students represented schools located in metropolitan areas (i.e., cities) of Australia while 38% were from regional or rural (i.e., country) areas. Hence, NVSES provided a critical opportunity for participation by students outside of major cities.

The target audience for NVSES was students already interested in and successful learners of science. As shown in Fig. 1, these students selected either 'Strongly

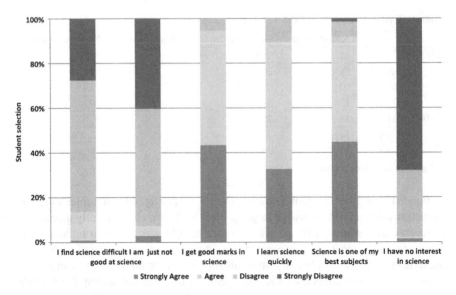

Fig. 1 Students' views of their ability in science (*N* = 150)

agree' or 'Agree' to the constructs *I get good marks in science* (94%), *I learn quickly in science* (90%) and *Science is one of my best subjects* (92%). Equally, they chose 'Strongly disagree' and 'Disagree' for the constructs *I find science difficult* (87%), *I am not good at science* (92%) and *I have no interest in science* (98%).

These data were derived from a PISA item (OECD, n.d.) included in the pre surveys that were completed with students prior to their engagement with NVSES. In another item, students were asked to identify the three careers they were interested in pursuing. The first identified career for students is summarised in Fig. 2. As viewed here, the three most frequently cited careers were engineering, medicine (as a MD), or a scientist. Even though 25% of the students indicated they were 'unsure' of their future careers, NVSES generally attracted and so aimed to enhance the opportunities for a select group of students who were already committed to and engaged with science thereby addressing Recommendation 1 from the *National Audit of Australian Engagement Activities* (Metcalfe, Alford, & Shore, 2013).

To capture students' reasons for participation in NVSES, they were asked in an open response item to provide their reasons for participation. As summarised in Fig. 3, the most prevalent reasons were to *increase understanding/learning/skills in science* (32%) followed closely by it is *different to school science* (23%). Other factors, such as interest in science, enjoyment of science and the chance to participate in NVSES were identified frequently in students' responses. It is important to note that 7% of students were compulsorily enrolled in NVSES by their teacher or school so their participation was not by choice. Not all students provided a reason for this item, and some identified more than one reason for participation.

Understanding the background of students, including their reasons for participation, was important information for the NVSES teachers but also for the project

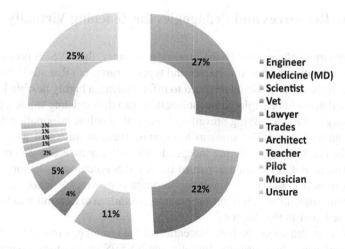

Fig. 2 Students' first-choice career interest as % of the total sample ($N = 150$)

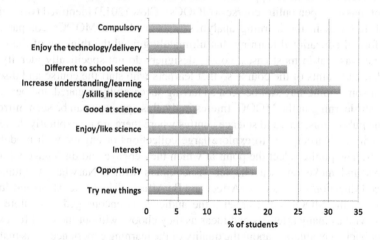

Fig. 3 Students' reasons for participating in NVSES ($N = 110$)

generally. Student interest and commitment to the virtual lessons was critical because, in the majority of cases, students were responsible for extricating themselves from their regular lessons and ensuring that they were connected into their NVSES lessons punctually. This required students to gain access to the virtual classroom via a link that was distributed to students in an email. However, the actual ways in which individuals or small groups of students connected into their lessons using WebEx® varied (see Appendix for details). Ultimately though, successful participation in NVSES relied on students being self-motivated with a degree of resilience.

Curating Resources and Pedagogies for Teaching Virtually

In creating virtual classrooms in the emerging sciences, the NVSES project team had to conceptualise the purpose, nature and types of resources that might be appropriate within the context. Given the need to offer students a highly flexible learning experience that could be explored in conjunction with their existing studies, consideration was given to identifying prevailing high quality online information or interactive resources for use. The aim was to organise these in ways that would allow students the flexibility of exploring independently online at their convenience, with collaborative activities included as part of the NVSES synchronous lessons. Where suitable resources could not be found, appropriate ones were designed, developed and used with students as part of the programme. Details as to how this was achieved are discussed later in the chapter.

With a clear idea about the broad intention of the teaching, a model for working with students was a high priority. Initially, the NVSES project team considered a self-directed autonomous approach similar to that offered by universities in the form of massive open online courses (MOOCs). Clow (2013) identified two significant differences in the learning analytics associated with MOOCs compared to more formal educational learning that ultimately guided the thinking of the team. The first was that in most cases, course designers do not specifically identify the intended end points of the course so that learners who start the course and discontinue at some point may still be seen as having successfully engaged and benefited from their learning in the MOOC. Interestingly, this approach can be seen mirrored in some public museums and science centres where there are no explicitly designed end-points for participants to explore large collections or engage with models or artefacts. The public select the point at which they engage and disengage with the displays and move on (see chapter "Encounters with a Narwhal: Revitalising Science Education's Capacity to Affect and Be Affected" by Steve Alsop and Justin Dillon). In the MOOC environment, the audience is encouraged to explore and engage with as many artefacts and ideas as they choose without the need for external feedback or evaluation about the quality of the learning experience. This manner of designing MOOCs and open learning repositories as a form of institutional outreach has shown considerable growth, while increasingly being marketed by higher education institutions (i.e., universities, vocational education) as demonstrating a branded commitment to broadening educational outreach and social responsibility (Gaebel, 2014; Macleod, Haywood, Woodgate, & Alkhatnai, 2015). An investigation of Mooc-list.com indicates that more than 80 course providers were offering more than 200 courses for study in 2015.

The second major difference between MOOCs and traditional models of educational learning is the recognised low course completion rates of students. Highly selective universities typically have completion rates above 90% and Open Universities and vocational colleges with broader social missions are typically above 60% (Clow, 2013). By contrast, most MOOCs have completion rates less than 10% with typically only 5–7% of participants finishing the course (Jordan,

2013). Surprisingly, data as to the reasons for learners discontinuing these courses are difficult to locate due to a lack of institutional sharing of creditable course analytics. Given the current available data, the NVSES design team decided that a more guided experience would benefit the younger Year 10 targeted audience as opposed to the open MOOC approach. The result was the inclusion of two synchronous lessons per week, which became pivotal components of the NVSES model supported by the online-curated resources.

An important consideration in the design of the synchronous lessons was varying the lesson format to help maintain student engagement through the units. As an example of lesson sequence, the Astrophysics unit encouraged students to explore and discuss a number of challenging 'big ideas' in a variety of ways. One such idea, astronomical scale, is particularly demanding because it is impossible to measure astronomical distances with familiar terrestrial units and it is difficult for most people to imagine the enormous distances between stars and galaxies. To help convey an appreciation of the universe's vast size students were initially asked to estimate a much smaller but more familiar distance, that is, the distance between the Earth and its Moon. Students were generally aware of the Apollo lunar landing missions and appreciate that they occurred some time ago (1969–1972). Most students assumed that given the limited technology available at that time the distance to the Moon is likely very much closer to Earth than it really is. Often this close proximity is reinforced as students encounter posters, textbook and internet images of the Earth-Moon system that are not drawn to scale so as to depict detailed surface features.

As an initial activity aimed at exploring the students' existing understandings, each student was asked to select two objects with a similar relative size to the Earth and Moon (e.g., a basketball and tennis ball) and to estimate their relative distance apart. Students then took pictures of their scaled model and uploaded them to a shared page on Padlet[2] in preparation for their next synchronous session. Padlet® enables invited student communities to post and share text messages, images, hyperlinks, and multimedia. During their next session students discussed their predictions and watched a short video by Derek Muller (Veritasium channel on YouTube), which focuses on a number of college students attempting a similar task using a basketball and an orange. Typically, most students were surprised to realise the true relative separation is about 30 Earth diameters and that in general people estimate the relative distance to be 10–20% of the true relative distance. This initial introductory activity to a small astronomical distance (Earth-Moon separation) of just 1.3 light seconds was then used to develop a concept of the 'light year' and parsec as useful large-scale units of astronomical distance.

Many of the lessons used purposefully designed presentations incorporating Prezi and PowerPoint, with hyperlinks to interactive learning tools (e.g., quizzes, reflective feedback, mind maps, and brainstorming activities) adapted for each lesson's objectives and context. Using this approach, the more engaging and visually informative websites and multimedia were introduced to students during the synchronous lessons. Students were then encouraged to explore these simulations and

---

[2] © Padlet.com registered to Wallwisher Inc.

interactive websites in their own time and with other students to assist in building their understanding of the key concepts. In addition, links to more complex websites and resources considered appropriate for advanced students to extend their understandings were provided for enrichment. Consistently, the approach adopted was "learner-focused" as supported by Rennie (2007) and Stocklmayer et al. (2010), as opposed to teacher-led and directed.

A key component of the NVSES design was curating electronic resources and activities (as referred to above) so they could be accessed by students to enhance their learning both in and beyond the virtual classroom. This required the design team to spend considerable time searching numerous scientific and higher educational websites to locate images, conceptual models and simulations that would enable students to explore key ideas and contentious issues related to the intended learning for each unit. Because the emerging sciences are developing so rapidly, careful attention was made to ensure that published information and resources were not just scientifically accurate but also contemporary in nature. In some units, such as Astrophysics and Nanotechnology, the number of web resources available on the Internet is almost overwhelming. However, on closer examination by the specialist NVSES teachers and academics (scientists and science educators working in collaboration) it became apparent that the scientific accuracy across these websites was often highly variable and that for many the cognitive level required to interpret the content was potentially problematic for Year 10 students.

Importantly though, the selection and curation of resources highlighted the difficulties that might be faced by many novice self-directed learners in attempting to use the Internet to explore the multitude of resources available without the necessary skills and specialist knowledge required to gauge the scientific accuracy of the information. This problem has been well known for some time. Research by Ng and Gunstone (2002) into students' perceptions of using the Internet to undertake scientific research (specifically the topic of photosynthesis) revealed how challenging 15-year-old students found this type of task to be. The small study ($N = 22$) involved mainly highly motivated and self-directed students. The findings identified several advantages associated with undertaking this type of learning: (i) access to almost unlimited information in a variety of formats, including text, images, and video— and this was 15 years ago; (ii) ease of access from any device at any time; and (iii) being able to access work at your own pace, using preferred technology and research methods. But there was also widespread agreement among the students about the difficulties they encountered. The majority of students reported that the information on most websites or video presentations was far too complex or technical to understand because it assumed a very high level of scientific content understanding that was beyond the novice learner. Students also acknowledged the length of time required to find just a small number of useful sites containing accurate information appropriate to the level they required. More surprising was the common agreement by students as to the importance of having access to a teacher with whom they could seek advice on a range of technological and content issues while undertaking their own research. As stated by two students in the article: "the teacher assisted me when I had difficulties and even showed me some websites that had the information I needed" and "the teacher put me back on the right track" (p. 497). These findings strengthened the case for regular synchronous teacher support and guidance throughout the implementation of the NVSES progamme.

Enriching Student Learning in Science

A key premise of the NVSES progamme was to encourage students to become active communicators, identify and question evidence, and debate contentious ideas presented to them rather than being passive consumers of scientific content. As such, many of the synchronous lessons included interactions with relevant research scientists to promote peer discussion in groups so that students from different schools and states could work collaboratively on shared tasks. Once the class groups were assigned, the teachers were able to drop 'virtually' in and out of these smaller groups as needed to provide support, listen to comments, and generally monitor student progress with learning tasks and activities. This group 'break out' feature was widely used to encourage engagement amongst students through collaborative tasks and to reduce distractions caused by off task 'chatter' that can occur with larger class numbers.

On occasions these break out groups were also used to allow teachers to gain insights into students' existing prior knowledge of particular related content. For example, the Nanoscience unit introduced an activity in lesson 3 (http://www.nvses. edu.au/lesson-plans/nanoscience/documents/Lesson-3.1-Metal-Lotus-diagram. docx) where students worked in break out groups to populate a blank lotus diagram with ideas related to eight key categories related to metals: alloys, uses, elements, properties, models, bonding, lattices and additional ideas. The intention was for students to discuss and add ideas related to as many categories as they could in 10 min while the teachers moved virtually from group to group seeking clarification about each of the additions contributed to the lotus diagram. The teachers reported that while this activity promoted strong student collaboration it also provided very useful feedback about the extent and depth of prior knowledge held by the students, with levels varying markedly between different cohorts of students.

Importantly, NVSES teachers were strongly committed to the inclusion of practical investigations. The difficulty was that traditional guided activities undertaken in face-to-face lessons could not be implemented without assistance from the participating schools, raising all kinds of issues for the project team. As an alternative, NVSES teachers identified a number of key practical activities that could be incorporated into lessons in ways that required everyday equipment and resources that were available in most homes while posing only low or no safety threats to students. When this alternative was not possible the practical activities were conducted live as teacher demonstrations during the lessons or pre-recorded for asynchronous access.

In some instances, students conducted investigations at home and returned to the next lesson with their findings and ideas to share with their peers. One such activity during the Nanotechnology unit in lesson 2, encouraged students to view a short YouTube video clip on how to prepare 'magic sand', a form of hydrophobic sand made from simple items found around the house (see http://www.nvses.edu.au/lesson-plans/nanotechnology/2.2.html#). Students first baked a small quantity of sand at low temperature for about an hour and then allowed it to cool. The sand was then spread evenly and sprayed with a hydrophobic silicon spray, for example, Scotch Guard and again allowed to dry. The sand behaves as expected when dry but when added to water it forms unusually shaped rigid structures without ever becoming wet. Students were keen to undertake this at home and report back on the curious discoveries about its unusual properties.

Supporting these practical activities was the use of simulations and modeling tools that allowed students to manipulate, isolate or vary individual variables so they could explore and better understand their impact on the system being studied. While some of these were already freely available through scientific websites, NVSES teachers identified additional activities they wanted to explore that did not have existing website equivalents. An Australian Ed-Tech company, Smart Sparrow, was employed to build several virtual laboratory investigations using their innovative Adaptive eLearning Platform (AeP). The state of the art, virtual laboratory activities were designed to allow students to explore the process of producing nanogold particles and the subsequent measurement of its physical properties. This laboratory activity is frequently undertaken in tertiary nanotechnology lessons. However, the high cost of the reagents required make it an expensive proposition for widespread use so that it becomes financially unviable for secondary schools. By working with educational programmers at Smart Sparrow, the NVSES project team developed several unique virtual laboratory investigations. Importantly, these authentic simulations offered multiple activity pathways for students to complete the tasks. They also provided teachers with access to personalised learning metrics for individual students as a means of monitoring individual progress through the activity. As shown in Fig. 4 using the example of 'nanogold', students were able to virtually manipulate images of laboratory equipment as they were guided via prompts to measure, combine and heat assorted reagents leading to the successful production of nanogold particles. Students then used a 'virtual' polarising laser and colour matching test to estimate the size of the nanoparticles produced. Finally, students used a sample of the nanogold particles produced to compare the relative strength of electrolytes present in several common sports drinks.

In addition to these specifically-developed virtual laboratory activities students were encouraged to explore a wide variety of engaging interactive computer simulations. These simulations were chosen for their high quality graphics and the opportunities they provided students to explore and identify features of their natural world that would normally be too small or too distant to appreciate. For example, helping students develop an understanding of black holes or the structure and nature of dark matter in Astrophysics could be readily supported in a virtual environment with the use of interactive simulations or multimedia.

To gain student perceptions of their learning through NVSES, feedback was provided from post-surveys with students. When asked: *What has been a highlight of*

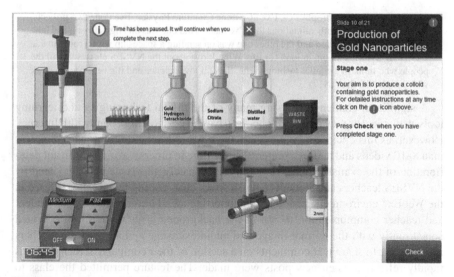

Fig. 4 Screen capture of the NVSES/Smart Sparrow collaboration – Nanogold

NVSES? students were positive about the NVSES learning experience, identifying the following key themes over a number of iterations of the emerging science units: (i) it was different to their school science experience; (ii) the interactivity and range of activities undertaken; (iii) it was relevant; (iv) opportunities to engage with students from other schools across Australia; (v) learning more about the subject area they had selected; (vi) working with other enthusiastic students and teachers who have expertise in the area; and (vii) the chance to talk with scientists and other experts in the field. These themes are exemplified in the following quotes from students and represent a credible cross-section of these data.

> I've really enjoyed the NVSES classes, they're a great learning experience and obviously what we learned was very different to what we learn in our normal school curriculum so it was very interesting and the online classroom was great as well, a very different way of learning. (S23)

> We got to learn some things that were outside of our normal school curriculum, and being able to interact with all the other students from across Australia in the online set-up was really great as well. Having different teachers with a different teaching style was also great. (S76)

> The overall learning experience has been a highlight this semester, because I was taught several new subjects, which definitely helped with my overall knowledge in science. (S34)

> Having the ability to participate in really interesting discussion and learning about mysterious and amazing concepts. (S12)

> It has showed me that the universe is made up of lots of interesting thing that are so tiny it's amazing! (S6)

I'm currently doing Astrophysics, I've also done Quantum Physics, I've done two projects, one for each, they're both animations that I've made. And one was on Higgs Boson and how it was recently discovered and the other one was on sending a spaceship with a robot inside it to Titan, Saturn's moon, to look for life. What I love about the NVSES classes are all the people who think in a similar fashion to me, they all love science and it's very cool to work with people like that. (S17)

A significant challenge to using many of the digital technologies and learning tools available was they required schools to have access to a high bandwidth Internet. However, as this could not be assumed, there was a limitation to broadcasting less than 8–10 videos and audio comments at one time in order to avoid significant deterioration of the available network quality. Fortunately, this issue was alleviated by the NVSES teachers and students using a 'back chat' texting feature also offered in the WebEx® environment as a way of supporting an additional channel for student and teacher communication to run in the background. This texting tool operated concurrently with the normal image and audio exchange and allowed the teachers and students to share text comments on a section of their computer screen that was rapidly refreshed when new posts were made. The feature permitted the class to exchange ideas in the background or to pose questions or comments to the teacher (or to each other) in ways that could be dealt with at appropriate times in the lesson. The teachers and students used back chat to circulate brief instructions containing hyperlinks to shared Google document folders or URL's for access to the many shared web-based resources (see Lancaster, Panizzon, & Corrigan, 2016). The back chat channel also allowed students to use a choice of emoticons. These were often used by the teachers to gain rapid feedback from the class about their levels of understanding, progress with an activity or the pace of the lesson. For example, 'tick' or 'cross' icons were posted by students to polls or quizzes set up by their teachers. It was also used successfully to provide positive feedback to the individual contributions made by students during the lesson.

In summary, the curated 'fit for purpose' resources were considered an integral component of the NVSES programme in supporting curious students to explore and extend their knowledge and understandings beyond the ideas introduced during the virtual lessons. Hence, the NVSES model might best be described as an outreach programme using a blended approach of virtual face-to-face lessons supported through scaffolded independent exploration making this 'active' outreach rather than the more commonly found passive experiences as highlighted by Metcalfe, Alford, & Shore (2013). The virtual lessons aimed to introduce content, generate discussion, provide support, and build and sustain student motivation. The additional curated resources encouraged students to explore and extend their understandings of content issues and their impact in their own time.

Challenges of NVSES As a Model for Future Outreach

The discussion of NVSES so far has focused on the positive aspects and outcomes but there were a number of critical challenges for the project team in implementing the programme over the 21-month period. These are included because much was learned in working around them that might be useful for similar programmes in the future.

Establishing and Building Rapport with Students: Relationships with students are core to quality teaching but it was the aspect that confounded the NVSES teachers most in the programme. When the first NVSES synchronous lesson was conducted students connected through WebEx® but refused to switch on the webcams or talk openly making it impossible for teachers to gain the indirect cues from the students easily achieved in face-to-face teaching. The result was that this introductory session became a lecture with most interaction occurring between the two NVSES teachers. After this session a number of decisions were made in preparation for the second lesson including making the turning on of webcams compulsory while allocating time at the beginning of the session to informally chat to students. In subsequent lessons teachers talked individually to students as they connected in so that all students used a webcam to facilitate a visual connection with the teachers. Several of the NVSES teachers reported using the connect time at the beginning and end of each synchronous session to encourage students to communicate with each other and their teachers more informally as well (e.g., via direct email). This provided important opportunities for the students and teachers to chat and help build digital skills and establish more productive relationships.

As each pair of NVSES teachers became more comfortable within the online environment and managing the technology, they actually found the back chat feature to be the most valuable way of providing immediate feedback to students, joke with students 'in the moment' and actively monitor student progress during a session. Representative examples of the comments made by NVSES teachers (four of the six) during interviews included:

> To be able to paste a link from the address in Google into the back chat which students can access straight away has been great—it expedites the process while taking nothing away from the teaching and interaction at the time—it sort of supports in the background what is happening. (T3)

> It is one of our best methods for monitoring student progress—what is being understood, where the questions are located, so what we might need to review in moving forward. (T2)

> Unlike the face-to-face lessons, this actually gives some insight into what students are thinking immediately as they write it down straight away—the immediacy is very different to my normal teaching. (T5)

> We have found it handy to chat with a student who may have been quiet for a while—you can just check they are ok and this helps to build that rapport with students. (T1)

While considerable progress was achieved throughout the NVSES programme, teachers still considered there was some way to go around developing further strategies for building relationships with students in the virtual environment.

Assumptions About Students and Digital Technologies: In conceiving and designing the NVSES model, there was an assumption by the principals and teachers from participating schools that their students were 'digital natives' so they would adapt to the virtual classroom easily. The NVSES teachers were also from a school that incorporated a high use of digital technologies in science lessons. As a result it was surprising when the first cohort of students appeared to struggle during the initial Astrophysics unit in Term 1, 2013—not with the scientific content but in dealing with the range of digital tools used in the teaching. In the post-surveys, the 26 students were asked: *Were there aspects about the technologies used during lessons that were challenging?*

> Technology is tricky—not having to open about four links just to get to the astroblog would be helpful. I found this really confusing. (S11)

> The websites could be easier to find—I really struggled understanding where to go from one place to the other. (S19)

> Having to move from the classroom to a ppt [powerpoint] and then somewhere else was confusing—I just got lost. (S31)

> Stick with one website so we don't have to travel into 6 billion different websites and make it easier to access and get better type of online environment to organise the teaching each lesson. (S87)

These types of comments were less frequent as the programme progressed, due to NVSES teachers reflecting on their teaching and what worked, thereby reducing student movement between 'spaces' and platforms in each lesson. So even though there was a view that navigating between the WebEx based-classroom, Google Docs and the Internet would be second nature to these digital learners, it was clear that the early adopters of the programme were uncomfortable with and frustrated by not knowing how to do this or not being directed about exactly what was required by their NVSES teacher. In order to address this oversight, the teachers spent the first synchronous session with every new cohort of students each term introducing them to the environment and helping to familiarise them with the various digital tools that would be used.

So an important point emerging for the project team was that while students might be comfortable with social media, participation in a virtual classroom adds a different layer of pressure and complexity that students are not necessarily familiar with, given there is a degree of accountability and the expectation that something will be achieved or learned at the end of the unit. In the case of NVSES, students did receive a report because the participating schools required this output. This component picks up an important difference highlighted by Stocklmayer et al. (2010), discussed earlier, where *learning* is the endpoint of an outreach activity as opposed to more traditional outcomes involving forms of accountability (i.e., completion of a report for assessment). When learning is the endpoint, the potential benefits that outreach opportunities provide in supporting and enhancing students' understanding of science and its processes as well as generating interest and engagement are increased. However, it comes at the expense of the student having more at stake when these conditions prevail.

Connecting to the Virtual Classroom: A reoccurring challenge for the NVSES teachers was the late arrival of students into lessons, which was in most cases due to students not allowing enough time to extricate themselves from lessons in their physical schools, move to a quiet location and connect into the NVSES session. The result was the NVSES session was interrupted frequently as students randomly connected into class, with teachers feeling the need to bring late students up-to-speed. Without a teacher from the participating school following up with students, it was often left to the NVSES project team to try and minimise this interruption by dealing with recalcitrant students located at a distance who were really under no obligation to attend NVSES lessons (i.e., unlike school, which was mandatory). This issue remained a challenge throughout the programme.

In relation to the students, many experienced considerable frustration logging into NVSES lessons for the first time. This was without doubt the most common issue that was dealt with by the project officer in the first two weeks of every term. Surprisingly, many of the 'new' students failed to follow fairly simple instructions that were outlined in a personal email identifying their NVSES email address and an initial password that could then be changed. Unfortunately, the majority of students tended to use their gmail or individual accounts and passwords, which were entered automatically or populated when using their own laptops or electronic devices. The result was a spike in the number of emails sent to the team by frustrated and irate students who were unable to log in to lessons. As a result, a support person was employed during these busy periods at the beginning of each term to deal with the students to ensure a quick and efficient log-in process.

The other issue experienced by some students related to the speed and quality of the Internet available in their schools. While considerable care was taken in curating electronic resources used with the NVSES science units, some students still identified problems with "lag" or "slowness". Data to support this statement were gained in the post-surveys where students were asked: *In terms of the technology, the aspects that could be improved in Astrophysics [Quantum physics/Nanoscience/Nanotechnology] were...*

I realise this is not an NVSES issue, but we need greater internet speeds to ensure smoother video streams. (S121)

Well the school I'm at could have provided a faster internet so we didn't get lagging and or computer/laptop to help with the lesson. (S89)

The screen kept looking pixelated to me so when demonstrations were happening I couldn't see it very well and sometimes the sound would cut-out and the screen would reload so that made it hard to learn or listen to guest speakers. (S67)

> The whole lesson on a computer is a pretty cool idea. It has made everything in this class interesting so far. My camera or video was always glitchy and that made things difficult with the presentations but that was from the school's end I think. (S17)

Instances where the connectivity dropped out or lagged for students often did not align with the identified broadband level of connectivity in the participating schools. It became evident that Internet speeds were random and issues difficult to predict. For example, students in some rural schools identified no issues with connectivity while students in large metropolitan schools in Victoria experienced frequent glitches with slowness during NVSES lessons. Yet, both schools had the same type and level of broadband. Unfortunately, this issue was beyond the scope of the project team but certainly impacted the overall satisfaction and enjoyment of the students.

Conclusion

Outreach activities provide an important role in raising public awareness of issues in science that society needs to engage with in terms of our future possibilities. Outreach activities that develop learning pedagogies lie along a continuum from the 'passive' (e.g., transmissive) activities identified in the *National Audit of Australian Engagement Activities* (Metcalfe, Alford, & Shore, 2013) to the more interactive engagement of students in vibrant classrooms. From this perspective outreach activities no longer need to sit outside 'formal' education. NVSES has provided an example of an outreach activity that is located somewhere along this continuum.

However, the conditions around the NVSES programme need to be examined against the criteria identified by Rennie (2007) and others. For example, the focus of the programme provided learning opportunities that relied on the intrinsic motivation of the learners. This was partly due to the voluntary nature of the programme so that students were in lessons with like-minded students who wanted to be involved in exploring science. A strong sense of identifying with other similar students was an important condition for the success of this outreach programme. Additionally, the programme opened up opportunities for extending the formal school curriculum in ways that explicitly identified links with the curriculum, but used very different contexts and approaches for exploring such ideas. While there was assessment associated with this programme, it was designed primarily to give learners some indication of their learning progress. Having said this, the assessment could be included by teachers in the more formal assessment processes within the students' mainstream schools. The learning within the NVSES programme was designed for Year 10 students (15–16 year olds) and so was quite homogeneous in nature. Critically, this did not seem to detract from the success of the programme.

As indicated above, not all the conditions for outreach activities as outlined by Rennie (2007) were met in the NVSES programme. However, what appears to be fundamentally important is the focus of learner-led experiences. In such situations the role of the teachers required the curation and presentation of resources in ways that allowed the learner-led experiences to 'unfold'. In closing, the NVSES programme provides some insights into the role that such outreach activities can play in mainstream education. Indeed this blended model of learning can be useful for future forms of active rather than passive outreach in which key ideas, understandings and resources can provide contexts for students to explore and interact with experts and peers for high impact and engagement in a comparative low cost learning environment.

Further information regarding the National Virtual School of Emerging Sciences is available from: http://www.nvses.edu.au/index.html

Appendix

As a requirement of enrolment in the progamme, participating teachers had to ensure that each student from their school had access to a computer, webcam, and headset (to avoid feedback) with a microphone for clarity of speech. However, what was interesting to monitor over the duration of the project were the various ways in which schools set up their students to link into the NVSES classroom. For example:

- School A connected students as a single group (with only some students on individual computers) to form a traditional classroom with the whole group joining the virtual classroom through a projected computer image. Student participation in this model required a student to move to the front of the class, with all communication viewed by the group as a collective. Hence, it was a physical class embedded within a virtual environment with the school physics teacher overseeing the group.
- School B set up students on individual computers so that they were able to work independently while in the same physical space; they were thus still able to share collectively while participating individually. The teacher overseeing the students participated (viewed) with the students, and then debriefed with students after each class even though she was not a science teacher.
- School C had students studying at a distance so they logged into the virtual classroom from home while their teacher did the same in order to participate in each class. In this situation, each student was physically separated from their peers and joined into the virtual classroom independently.

In the majority of NVSES lessons, students connected in the same way as School B with students individually positioned on a computer using a headset. From the feedback gained from students and their teachers, few schools provided a designated teacher to support the students.

Acknowledgement NVSES would not have been possible without funding provided by the Australian Government Department of Education through the Broadband Enabled Education and Skills Services Programme.

References

Braund, M., & Reiss, M. (2006). Validity and worth in the science curriculum: Learning school science outside the laboratory. *The Curriculum Journal, 17*(3), 213–228.

Clow, D. (2013). MOOCs and the funnel of participation. In D. Suthers, K. Verbert, E. Duval, & X. Ochoa (Eds.), *Proceedings of the third international conference on learning analytics and knowledge* (pp. 185–189). New York, NY: Association for Computing Machinery.

Falk, J. H., & Dierking, L. D. (2000). *Learning from museums.* Walnut Creek, CA: AltaMira.

Gaebel, M. (2014). *MOOCs Massive Open Online Courses—An update of EUA's first paper* (EUA Occasional Papers). Retrieved from http://www.eua.be/Libraries/publication/ MOOCs_Update_January_2014

Hodson, D. (1998). Is this really what scientists do? Seeking a more authentic science in and beyond the school laboratory. In J. Wellington (Ed.), *Practical work in school science. Which way now?* (pp. 93–108). London: Routledge.

Jordan, K. (2013). *MOOC completion rates: The data.* Retrieved from http://www.katyjordan. com/MOOCproject.html

Lancaster, G., Panizzon, D., & Corrigan, D. (2016). Pursuing different forms of science learning through innovative curriculum implementation. In D. Corrigan, C. Buntting, J. Dillon, R. Gunstone, & A. Jones (Eds.), *The future in learning science: what's in it for the learner?* (pp. 101–126). New York, NY: Springer.

Macleod, H., Haywood, J., Woodgate, A., & Alkhatnai, M. (2015). Emerging patterns in MOOCs: Learners, course designs and directions. *TechTrends, 59*(1), 56–63.

Malcolm, J., Hodkinson, P., & Colley, H. (2003). The interrelationships between informal and formal learning. *Journal of Workplace Learning, 15*(7/8), 313–318.

Metcalfe, J., Alford, K., & Shore, J. (2013). *National audit of Australian science engagement activities 2012. Inspiring Australia—A national strategy for engagement with the sciences program.* Retrieved from http://www.econnect.com.au/wp-content/uploads/2013/01/National-Audit-Final-Report_20-01-13-FINAL.pdf

Ng, W., & Gunstone, R. (2002). Students' perceptions of the effectiveness of the world wide web as a research and teaching tool in science learning. *Research in Science Education, 32*(4), 489–510.

OECD. (n.d.). *PISA 2006. Released items—Science.* Retrieved from https://www.acer.edu.au/files/ pisa_relitems_sc_2006ms_eng1.pdf

Rennie, L. (2007). Learning science outside of school. In S. K. Abell & N. G. Lederman (Eds.), *Handbook of research on science education* (pp. 125–167). Mahwah, NJ: Lawrence Erlbaum.

Stocklmayer, M., Rennie, L. J., & Gilbert, J. K. (2010). The roles of the formal and informal sectors in the provision of effective science education. *Studies in Science Education, 46*(1), 1–44.

Using a Digital Platform to Mediate Intentional and Incidental Science Learning

Cathy Buntting, Alister Jones, and Bronwen Cowie

Abstract This chapter challenges the distinction between informal and formal science learning in the context of learning science from online resources, arguing instead for consideration of intentional and incidental science learning. The New Zealand Science Learning Hub (sciencelearn.org.nz) is used as an example to demonstrate how both intentional and incidental learning can be supported. The aim of the Hub, which is Government-funded, is to make contemporary science research and development more accessible to teachers, students and the wider community through presenting the stories of science and scientists in multimedia format, supported by resources for teaching and learning. To foster ongoing engagement of teachers, as a key target audience, the Hub has a deliberate strategy to support both intentional and incidental learning through website design and through connecting with topical science-related events and news. The embedded social media strategy in particular is a tool that mediates incidental engagement with the Hub content.

Keywords Intentional learning · Incidental learning · Digital platforms

Introduction

Several chapters in this book explore how the advent and expansion of the Internet, and an increasing array of 'smart' digital devices, have resulted in unprecedented access to scientific information—and multiple views about this information—changing when, where and how science learning can take place. In addition, "The changing nature of the web as well as the changing nature of 'classrooms' where learning can take place across physical and cyber spaces in and out of school provides learners with an array of choices for the topic and location of their learning experiences" (Baram-Tsabari, 2015, p. 1109). Within this context of a changing learning landscape, this chapter introduces a consideration of intentional and

C. Buntting (✉) · A. Jones · B. Cowie
University of Waikato, Hamilton, New Zealand
e-mail: BUNTTING@waikato.ac.nz; a.jones@waikato.ac.nz; bronwen.cowie@waikato.ac.nz

incidental learning when designing an on-line resource that supports science learning, whether in or out of school, and whether by students, teachers, or other communities.

While the boundaries between intentional and incidental learning are permeable and fluid, in this chapter we define intentional learning as learning in response to a specific, purpose-driven inquiry. In other words, intentional learning is learning that occurs when the learner actively sets out to learn something specific. On the other hand, incidental learning is unintentional or unplanned learning (Kerka, 2000), and often "a by-product of some other activity, such as task accomplishment, interpersonal interaction, sensing the organizational [or school] culture, trial-and-error experimentation, or even formal learning" (Marsick & Watkins, 1990, p. 12). Other examples of episodes triggering incidental learning include reading pertinent materials, or observing peers, supervisors and 'veterans' (Marsick, Watkins, Callahan, & Volpe, 2006). Importantly, because incidental learning is often not recognised or labelled as learning, it can be difficult to measure. It therefore is less likely to be communicated to others (Matlay, 2000). It should be noted, however, that both incidental and intentional learning tend to be situated, contextual, and social. In addition, incidental learning often leads to episodes of intentional learning.

Take a school student who is searching the Internet for information on a given topic for the purposes of a school assignment—processes involved in xenotransplantation, say. Along the way, the student becomes embroiled in online discussion about the use of animals for testing cosmetics. While learning about the processes involved in xenotransplantation reflects intentional learning, finding out about some of the ethical issues associated with these processes may have been incidental (depending on the parameters of the student's assignment), as was her subsequent engagement in an online forum debating the use of animal models in cosmetic testing. Or imagine a teacher looking for resources to help him scaffold his students' understanding of photosynthesis, and strategies to help the students unpack their various alternative conceptions about this process. Imagine, too, that along the way the teacher finds out about research into the potential effects of climate change on photosynthesis in the oceans, and the impacts on multiple food webs and atmospheric oxygen levels. This is not learning that he was intentionally seeking; it happened incidentally, as a by-product of his more directed (intentional) learning endeavour.

With these examples in mind, we argue that when designing web-mediated learning experiences, it is worth considering intentional and incidental learning— and how the two can mutually support each other. Our intention in this chapter is therefore to consider the roles of both incidental and intentional learning when users engage with an online science education resource, and how both types of learning can be supported in order to foster their ongoing engagement with the resource. The example that we use is the New Zealand Science Learning Hub (sciencelearn.org. nz). The aim of this resource, which is Government-funded, is to make contemporary science research and development more accessible to teachers, students and the wider community through presenting the stories of science and scientists in multimedia format, alongside resources to support teaching and learning.

The Role of Digital Technologies in Supporting Intentional and Incidental Learning

The rise of digital technologies, including desktop and laptop computers, and a wide array of mobile devices, has shifted the ways individuals can engage in both intentional and incidental learning. The Internet, for example, offers ready access to a vast corpus of knowledge, often presented in multimodal form. While many websites are designed with the specific needs of potential visitors in mind, a case can easily be made for ways in which they also support incidental learning. In addition, Web 2.0 technologies, such as wikis, blogs, vlogs, podcasts, and social networking sites, offer new capacities for seeking, sharing and curating information, and enable interested participants to contribute in an active and dynamic way. Discussion forums and even the comments sections at the ends of articles enable wide participation, and numerous opportunities for both intentional and incidental learning. Even two decades ago, when the Internet was still in its infancy, online discussion group members were articulating intentions to learn, as well as recognising instances of incidental learning—over half the participants in a study by Collins and Berge (1996) reported learning both incidentally and intentionally at different times.

Digital platforms therefore offer compelling opportunities to link intentional and incidental science learning, in both formal and informal learning contexts. For instance, Baram-Tsabari (2015) reports that:

> Studies measuring public interests in science have found that searches for general and well-established science terms were strongly linked to the academic calendar, meaning that the trigger for the search was probably the education system. On the other hand, searches for concepts related to ad hoc events (e.g., Nobel Prize announcements) and current concerns were better aligned with media coverage. (p. 1108)

Baram-Tsabari also reports that analysis of reader comments to online news articles indicates that the most fruitful discussions are initiated in the discussion threads themselves, rather than in the science news articles. These discussions offer fruitful avenues for incidental learning. Shanahan (2015) further argues that it is these online interactions that are a crucial aspect of responsible citizenship that today's students need to be equipped to engage in. She describes how digital technologies have changed public access to new scientific developments, or "science-in-the-making"—scientific findings, press releases, media reports, public and professional commentaries, and often the researchers themselves are now publicly accessible via the Internet. In addition, debates that might hitherto have occurred behind the closed doors of the science academy are now publicly accessible, which can be unsettling and confusing for non-scientists. Here, interesting issues arise: if a reader looking for information about a specific scientific phenomenon stumbles across information that is apparently being hotly debated, what might the incidental learning be with respect to the reader's views of the nature of scientific development?

In addition, search results depend significantly on the search terms that are used—but they are also often prioritised according to algorithms and audience metrics. Brossard and Scheufele (2013) report on studies that show

> clear discrepancies between what people search for online, which specific areas are suggested to them by search engines, and what people ultimately find. As a result, someone's initial question about a scientific topic, the search results offered by a search engine, and the algorithms that a search provider uses to tailor retrieved content to a search may all be linked in a self-reinforcing informational spiral in which search queries and the resulting Web traffic drive algorithms and vice versa. (p. 41)

This is important, and reinforces how search results may shape users' perceptions, knowledge, and discourse about emerging technologies (Baram-Tsabari, 2015).

Further, there are indications that online users—at least in the U.S.—are turning more and more to blogs and other information available only online, and less to online versions of traditional news outlets (National Science Board, 2014). Again, what might the incidental learning be if the reader is not a critical consumer of the information that they are using? For example, Anderson, Brossard, Scheufele, Xenos, and Ladwig (2014) investigated the impact of comments following a news item on nanotechnology as an emerging technology, finding that the tone impacted significantly on readers' interpretations of potential risks associated with the technology—even though the comments were consistent in terms of content.

At a time when the Internet has become the main source of science-related information for Western societies (National Science Board, 2014), and when teachers of science are increasingly called to help students become critical consumers—or connoisseurs (Fensham, 2015)—of science, web-based opportunities for both intentional and incidental learning, whether in informal or formal contexts, are worth exploring. The remainder of this chapter focuses on the provision of opportunities for both intentional and incidental learning in order to foster ongoing engagement with an online resource developed to support science teaching and learning, the New Zealand Science Learning Hub.

The Science Learning Hub

The Science Learning Hub (sciencelearn.org.nz) is an online resource funded by the New Zealand Government since 2006. Although its target audience was initially New Zealand teachers, Google Analytics and annual surveys indicate that it is accessed by a far broader audience, including school and university students, parents, scientists, and the general public. Significant traffic outside of New Zealand comes from the U.S., U.K., Australia, India, Philippines, Canada, Spain, Malaysia and Pakistan. The programme of work since the resource was launched has focused on the development of content, promotion among teachers, and professional development for teachers.

The purpose of the Hub is to make contemporary New Zealand science research and development more accessible to a school audience by enabling teachers to keep up to date with scientific developments, and supporting them to develop engaging teaching and learning programmes that use these developments as contexts for science teaching and learning. It is funded by the New Zealand Government through Vote Science, rather than Vote Education, and was initiated as a mechanism to enable science organisations to communicate in a sustainable manner with New Zealand's school audience and the wider community. It currently forms a key part of the Government's 'science in society' strategy (New Zealand Government, 2014). Importantly, in addition to science content, users are invited to engage with ethical aspects of scientific and technological advances—values development is an important aspect underpinning *The New Zealand Curriculum* (Ministry of Education, 2007).

Key to the ongoing success of the Hub has been the provision of quality-assured resources developed by a team that includes teachers, science education researchers, scientists, and multimedia developers. Typically, a topic or theme for a collection of resources is selected; scientists working in the area are identified and approached to participate in the project; a writer works with the scientists to identify the key science ideas, and then the writer translates these into text for the Hub; multimedia content such as short video clips, interactives and animations are developed; relevant curriculum links are identified and teaching resources developed; and opportunities for linking to existing Hub content are identified. In addition, the Hub leverages off other science communication initiatives, such as science-related television shows, radio broadcasts and education events such as 'Sea Week', providing significant value-add to resources developed through other funding streams and initiatives. More recently, the advent of social media has been leveraged to support ongoing engagement of teachers and other audiences with the Hub.

Teachers, students and other users can and do engage with the content of the Hub in a variety of ways, including by accessing and downloading the resources, and by interacting directly with members of the Hub team and/or other Hub users. Direct interactions can occur through face-to-face and online professional development sessions or direct contact (phone or email); or through social media activity. While some interaction with other Hub users takes place during the professional development sessions, it is the social media environment that particularly supports access to a large number of other (and diverse) users, including teachers, teacher educators, scientists, science communicators, parents, and other interested people. Key to the Hub's ongoing engagement strategy is the provision of both intentional and incidental learning opportunities, considered below.

Design Features to Support Intentional Learning

An onsite survey, administered annually since 2010, has consistently highlighted that New Zealand teachers visit the Hub primarily to find content for students and ideas for teaching. A large proportion also access the Hub to find content for themselves, and for general interest. In many of these cases, the visits would likely have started with intentional purposes: to find specific content, for their students or themselves. Because of this, the provision of intuitive navigation and search strategies is integral to the Hub's structure and functioning.

With a vast smorgasbord of resources, including over 5500 written pages, 3000 images, 1000 videos, 800 teacher resources, and nearly 100 interactives and animations, our strategy for supporting intentional learning focuses on highlighting key content collections, and extensive metadata tags embedded across the content. This tagging feeds into the advanced search engine, which enables users to search by topics, curriculum strands, year levels, and resource type. It also means that the site is directly searchable from other NZ-based sites with whom we have strategic relationships, including Te Kete Ipurangi (a Government-funded portal for the education sector), Pond (a recently-released Government-funded teacher curation and collaboration tool), and DigitalNZ (led by the National Library, with the intention of making NZ digital content more useful). Once a page is selected, the top and side navigation bars enable users to easily identify related content—although this can be where the lines between intentional and incidental learning start to blur, as the reader is diverted by interesting tangents, exploring content that piques their interest. Indeed, the tensions between finding what you are looking for, and pursuing interesting diversions, are a consideration for all web design.

Being closely linked to *The New Zealand Curriculum* (Ministry of Education, 2007) is a second key aspect in our strategy to support the intentional learning of teacher visitors to the site. *The New Zealand Curriculum* strongly advocates for context-based approaches to teaching and learning in general, and in the science learning area emphasises the importance and relevance of science to everyday life. In line with this, New Zealand teachers using the Hub consistently report they particularly value the New Zealand-based examples (Chen & Cowie, 2013). For example, responses to the 2015 open-ended questions included comments such as:

> Your site provides a wonderful look at space but also the world of science through the New Zealand science community.

> The students enjoyed the resources and the NZ based, up to the minute current content is unique. I have found it particularly helpful to support teaching of the nature of science strand, which can be difficult to find material to support.

> [I used] NMR videos with level 3 chemistry topic of spectroscopy, which allowed students to see how this applied to the real world of science research. The students were more motivated as they could see why they were learning about spectroscopy.

> Last year during a unit focusing on adaptation I used the Marine Ecosystem interactive as a teaching tool. This was highly engaging for students and allowed them to make links between their existing knowledge and a real life simulation.

An extensive programme of classroom-based research supports survey evidence that the videos are particularly valued for the ways they address the intended learning objectives identified by the teacher. These 'short and sharp' clips—over 50 h' worth, broken into 2–4 min clips—feature scientists talking about their own cutting-edge research and its applications in society, along with their personal stories about what they love about their work, the challenges of science, and how and why they became scientists. Our research suggests that these 'virtual' scientists visiting classrooms via a video recording can achieve many of the same benefits as scientists actually visiting classrooms in person (Chen & Cowie, 2014). For example, watching scientists talk about their work challenged and expanded students' views of who could be a scientist and what scientists do. Student commentary also indicated that the impact of seeing and hearing scientists talk about their work was amplified because they were New Zealand scientists talking about how their scientific work contributed to the well-being of local communities and the environment. A second benefit of the videos is that they support students' and teachers' science learning. Teachers reported that they learnt new subject content knowledge and specialised terminologies, how to pronounce science-specific words, and ideas about the nature of science from watching the videos. A third benefit reported by the teachers in the Chen and Cowie study was that using a video of a scientist talking allowed the teacher to step back from the role of classroom authority and focus on fostering discussion and debate, especially of more controversial ideas and practices. Furthermore, the short length of the videos meant that they could readily be integrated into a lesson as and when they would best contribute, and they could be revisited and reviewed—fostering both intentional and, at times, incidental learning opportunities.

Design Features to Support Incidental Learning

Incidental learning was described earlier as learning that is unplanned or unintended, and that occurs as a by-product of some other activity, including more intentional learning episodes. Of course, it is unclear from Google Analytics data whether users accessing the Hub either directly or via a search engine are looking for something specific (indicative of intentional learning), or browsing for ideas in general. For example, data for the period 1 January—30 June 2013 showed that over 63,000 New Zealand visitors came from Google or other search engines, a further 3000 from other websites, and 21,000 went directly to the site. For New Zealand teacher respondents to the 2015 on-site survey, and who used the Hub materials in the classroom, top sources of finding out about the Hub were science textbooks or teacher magazines (24%); formal professional development, including teacher conferences (22%); a search engine (22%); word of mouth (19%); and the Hub newsletter (8%).

A key purpose of website landing pages is to keep visitors on the site and, often, to alert visitors to the wide range of content that is available. To showcase the Hub's diverse content collections, a moving carousel of images was initially used, the images representing key collections, for example, Dating the Past, Fighting Infection, and Icy Ecosystems (see Fig. 1). The intention was to ensure that even

Fig. 1 An image of the pre-2015 home page, showcasing key content collections using a moving carousel

when visitors came to the site with a goal in mind, the carousel would offer them an appealing overview of the wide range of content that was available—and draw them into browsing through additional content. Over time, growth in the content meant that this carousel became over-crowded, and it was replaced with a rotating slideshow of pre-selected content considered to be of interest to a wide audience. The home page also featured a 'Spotlight' window and 'Science Teaching Ideas' window. Each of these features was regularly updated, with newly published resources, resources connected to a topical event, or feedback from a user highlighting how a particular resource has been used. In a further iteration, popular collections are highlighted and content is grouped into science topics and concepts that can be browsed or searched. Within the individual pages, side navigation bars again enable viewers to identify related content. This design is intended to support teachers (and other users) to scope the breadth of ideas that might be relevant to a particular science lesson or unit.

The iterative and ongoing design process highlights the importance—and challenge—of keeping pace with developments in website programming capabilities and user behaviours and expectations. The premise, however, is that pages are designed so that even when visitors are searching for a particular topic or teaching resource, their attention may be captured by additional ideas and routes to explore.

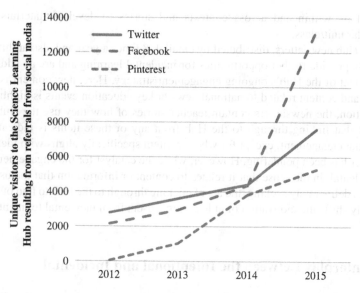

Fig. 2 Transferrals to the science learning hub from three social media platforms: Twitter, Facebook and Pinterest

Of additional significance to discussion about how incidental engagement and learning is promoted, visits to the site from social media sites are increasing (see Fig. 2). It seems reasonable to suggest that, in many cases, this catalyses incidental engagement and learning—content that is relatively unpredictable (from the user's perspective) shows up in users' news feeds across these social media platforms, and users clicking from these posts through to the Hub are likely doing so because something in the post was sufficiently interesting that they wanted to find out more. Of course, in some cases the incidental engagement likely turns into intentional learning, in the sense that the learner begins to seek something specific.

Social media has become a key part of the Hub's work programme, with a dedicated part-time social media expert who engages regularly across multiple platforms, and who has intimate knowledge of the Hub so as to rapidly identify relevant content in response to topical events such as national news items or events such as 'Primary Science Week'. Such activity has the potential to offer 'reform-minded teachers' the support they need to develop their practice (Goodyear, Casey, & Kirk, 2014). Additionally, much of the power of social media lies in the multiplier effect, with all followers able to tap into a post. Thus, while a social media action may be used to support intentional learning, for example, through collating new Pinterest boards in response to a teacher's request, its power lies in its multiplier effect. By way of a specific example, a board on the ebola virus, created in response to one teacher doing some late-night planning on this topic at the peak of media attention around the nature, spread and potential worldwide impact of ebola, rapidly had over 500 followers. Further highlighting the potential for incidental learning, curated collections such as Pinterest boards illustrate some of the ways teachers might look within and across contexts and stories to develop materials for a particular unit/class.

The Hub newsletters, distributed by email 4–8 times a year and also archived on the Hub, provide further opportunities for incidental learning and are an additional component of the Hub's ongoing engagement strategy. Here, new content is introduced, and content related to national news or key education events is highlighted. In addition, the newsletters contain teacher stories of how they use the Hubs. We suggest that linking through to the Hub from any of these items is indicative of incidental engagement, except for where an item specifically aligns with ideas that a teacher has been pondering. However, while the catalyst for the engagement may be incidental, in the sense that it relates to content or information that the user had not been deliberately seeking, the act of clicking through to the Hub to find out more probably shifts the motivation of at least some users from incidental to intentional.

The Interplay Between the Intentional and Incidental

In developing this chapter we began asking when incidental learning becomes intentional, and vice versa—and why this might matter? For example, being on social media is in the first instance an intentional action, but the socially interactive nature of this forum means that, once there, many opportunities exist for incidental learning. Further, if one comes across a tweet or Facebook post about the Hub, this information is serendipitous—but then the action to follow-up is deliberate, or intentional. Incidental learning in this case initiates ongoing intentional learning. It seems to us, therefore, that it is most likely that incidental and intentional learning are often in a dynamic relationship—and that both can be leveraged to foster ongoing engagement with online resources.

In the case of the Science Learning Hub, the work programme seeks to promote multiple routes to the Hub for teachers so that they might find it, and the ideas and resources embedded within it, intentionally *and* incidentally. Specifically considering both these approaches has been fundamental to our engagement strategy. In addition, one can distinguish between learning (whether intentional or incidental) that relates to learning about the science; learning about teaching the science; learning about making effective use of online technologies for ongoing professional support and inspiration; and even learning more about the Science Learning Hub, including how to access (whether by search or browsing) the extensive array of content, and the affordances provided by the multiple different types of content.

The professional development (PD) activities associated with and funded by the Hub play an important role in supporting teacher learning. Of course, while engagement in professional learning may be intentional, the reasons for engagement vary. For example, we know from a question all participants are asked when signing up for online PD sessions that many have not previously visited the Hub. It is reasonable to assume that for these participants, an intentional goal for their learning is to find out more about the Hub and how they might use it to support their science teaching and learning programmes. However, they may also have signed up for a particular PD session precisely because of the topic being covered, for example, finding the nature of science in a teaching topic, or using the Science Learning Hub to support teaching about the carbon cycle. It is conceivable, therefore, that those with the intention of learning more about the Hub will also learn more about a particular area of science (incidental learning), and vice versa. Indeed, teacher feedback to both online and face-to-face workshops indicates that they gain insights into the range of material on the Hub *and* guidance about how they might use the Hub, including to access specific content. At the same time, the opportunity and time to explore and discuss material that captures their attention affords incidental learning that can be leveraged later, when the occasion arises. Importantly, survey feedback has shown that teachers who have taken part in formal professional development are more likely to access and use Hub materials in their classrooms, suggesting that informal and incidental learning may not be sufficient for teachers to more fully engage with Hub resources. This does not seem unreasonable given the scope and depth of the material that is available.

Closing Thoughts

This chapter uses the Science Learning Hub as an example of a large educational resource specifically developed to support science learning—by teachers, students, and the wider community. In this context, the distinction between 'formal' and 'informal' learning seems less helpful—online resources can be accessed at any time, from anywhere, and for multiple different purposes. In the case of the Hub, we know that teachers visit the site for a variety of reasons, and that these are more or less tightly focused: some are searching specifically for information about an idea; others are looking for an idea, any idea; still others are seeking teaching and learning resources. We have therefore attempted to design the Hub so that learning opportunities are not limited to those that are intentionally initiated. We also understand that to maximise intentional learning, we need to tap into incidental learning and extend visitor interest and understanding beyond what might have been their immediate goal.

As with any initiative, issues of scale-up need to be considered. For us, this relates to both creating wider national engagement as well as creating locally-relevant content in an international context. In both cases, strategic website design and quality-assured content development needs to be complemented by ongoing activities that both raise the profile of the project and support teachers (as a key target audience) to enhance their uptake and adaptation of the vast array of resources. As the digital landscape continues to expand and change, we need to continually evolve in order to remain relevant and useful.

Acknowledgements We wish to acknowledge the Ministry of Business, Innovation and Employment for their work promoting broader social engagement in science, and for funding the Science Learning Hub. Mira Peter provided much of the quantitative data used in this chapter, and Cath Battersby and Rachel Douglas have taught us a lot about the value of social media for broadening and deepening teacher engagement in the Hub. The ongoing success of the Science Learning Hub depends on the diverse skills of the development team, the generosity of the science sector, and the commitment of teachers and others to engage deeply in contemporary science research.

References

Anderson, A. A., Brossard, D., Scheufele, D. A., Xenos, M. A., & Ladwig, P. (2014). The "nasty effect": Online incivility and risk perceptions of emerging technologies. *Journal of Computer-Mediated Communication, 19*, 373–387.

Baram-Tsabari, A. (2015). Web 2.0 resources for science education. In R. Gunstone (Ed.), *Encyclopedia of science education* (pp. 1107–1109). Dordrecht, The Netherlands: Springer.

Brossard, D., & Scheufele, D. A. (2013). Science, new media, and the public. *Science, 339*(6115), 40–41.

Chen, J., & Cowie, B. (2013). Engaging primary students in learning about New Zealand birds: A socially relevant context. *International Journal of Science Education, 35*(8), 1344–1366.

Chen, J., & Cowie, B. (2014). Scientists talking to students through videos. *International Journal of Science and Mathematics Education, 12*, 445–465.

Collins, M. P., & Berge, Z. L. (1996, October). *Mailing lists as a venue for adult learning.* Paper presented at the Eastern Adult, continuing and distance education conference, University Park, Pennsylvania.

Fensham, P. (2015). Connoisseurs of science: A next goal for science education? In D. Corrigan, C. Buntting, J. Dillon, A. Jones, & R. Gunstone (Eds.), *The future in learning science: What's in it for students?* (pp. 35–59). Dordrecht, The Netherlands: Springer.

Goodyear, V., Casey, A., & Kirk, D. (2014). Tweet me, message me, like me: Using social media to facilitate pedagogical change within an emerging community of practice. *Sport, Education and Society, 19*(70), 927–943.

Kerka, S. (2000). Incidental learning. In *Trends and issues alert, 18*. Columbus, OH: ERIC Clearinghouse on Adult, Career, and Vocational Education. Retrieved from http://www.eric.ed.gov/PDFS/ED446234.pdf

Marsick, V. J., & Watkins, K. (1990). *Informal and incidental learning in the workplace.* London, UK/New York, NY: Routledge.

Marsick, V. J., Watkins, K. E., Callahan, M. W., & Volpe, M. (2006). *Reviewing theory and research on informal and incidental learning.* Retrieved from http://eric.ed.gov/?id=ED492754

Matlay, H. (2000). Organisation learning in small learning organisations. *Education and Training, 42*(4–5), 202–210.

Ministry of Education. (2007). *The New Zealand curriculum.* Wellington, New Zealand: Learning Media.

National Science Board. (2014). *Science and technology: Public attitudes and understanding* (pp. 14–01). Arlington, VA: National Science Foundation (NSB).

New Zealand Government. (2014). A nation of curious minds—He whenua hihiri i te mahara. In *A national strategic plan for science in society.* Wellington, New Zealand: Government.

Shanahan, M.-C. (2015). When science changes: The impact of ICTs on preparing students for science outside of school. In D. Corrigan, C. Buntting, J. Dillon, A. Jones, & R. Gunstone (Eds.), *The future in learning science: What's in it for students?* (pp. 61–81). Dordrecht, The Netherlands: Springer.

"Meet the Scientist": How Pre-service Teachers Constructed Knowledge and Identities

Gillian Kidman and Karen Marangio

Abstract A continuing issue for tertiary educators, and pre-service teachers alike, is the articulation between university classes where the pre-service teacher is the user of knowledge, and the school setting where the pre-service teacher is the imparter of knowledge. It is not clear how easily pre-service teachers can transfer university learnings into 'in school' practice whilst on a practicum placement or as a beginning teacher. Similarly, it is not clear how easily knowledge, both contextual content and pedagogical knowledge, learned in the school can be dis-embedded from the particular school context and understood more generally by the pre-service teacher. The school and university settings demand different tools, social interactions and knowledges, and often contradictions occur. As problems arise, the pre-service teacher is required to integrate numerous elements from both contexts to provide a solution to the challenge. Pre-service teachers must not simply engage in a single setting at any one time, they must engage in multi-tasking within a single context, but also in multiple communities of practice simultaneously (Tsui, 2003). By integrating elements in multiple contexts, to solve problems, new learning occurs through the blend of ideas.

This proposed chapter explores the problems of transfer between the university setting and school setting for pre-service secondary science teachers. The chapter will explore the following research questions: (1) What informal science activity systems can be at play in school-university activity systems that impact upon the knowledge gain and transfer of preservice science teachers? (2) Are these activities sources of two-way knowledge transfer?

Keywords Pre-service teachers · University-school partnerships · Transfer of learning · Experiential learning · Meet a scientist

G. Kidman (✉) · K. Marangio
Monash University, Clayton, VIC, Australia
e-mail: gillian.kidman@monash.edu; karen.marangio@monash.edu

© Springer International Publishing AG, part of Springer Nature 2018
D. Corrigan et al. (eds.), *Navigating the Changing Landscape of Formal and Informal Science Learning Opportunities*,
https://doi.org/10.1007/978-3-319-89761-5_11

This chapter uses the framework of experiential learning to explore the informal learning resulting from a school-university-university partnership. The chapter explores how pre-service teachers constructed knowledge through informal learning interactions, especially a "meet the scientist" interaction boosted through the effects of a team context. Specifically, a group of Australian pre-service physics teachers planned a Year 12 physics lesson and a public seminar that required the pre-service teachers to meet scientists based in a university in the United States of America (US). The chapter begins with a contextual vignette.

The Story of Six Physics Pre-service Teachers "Meeting the Scientist"

This study originates in the realm of pre-service teacher education and the need for the assessment of the pre-service teachers' learnings. In a particular curriculum methods unit, the assessment task included a 30-min group presentation to peers in their tutorial group. Past offerings of the unit indicated the pre-service teachers did not enjoy peer presentations that had the audience 'pretending' to be 15-year-old students. To increase the authenticity of the task, and to increase audience enjoyment and learning, it became a requirement that the presentation take the form of a professional development (PD) hands-on session. A group of six pre-service physics teachers (PSPTs) were a subset of 95 pre-service secondary science teachers undertaking a 9 week curriculum methods unit at a large Australian university. The curriculum unit was the last of three core methods units; the focus of the unit was on laboratory-based pedagogies. The PSPTs were given the broad topic of 'nanotechnology' (nanotechnology is a topic that appears in the senior science curriculum for students in their final 2 years of non-compulsory education in Australia, and was likely to be outside the experiences of the PSPTs). The PSPTs were also provided with the e-mail address of a scientist at a US University. This particular scientist, Helen,[1] was a research scientist who also worked in a team providing PD to local high school teachers. The first author was aware of the PD, and had arranged with Helen to have email conversation with the Australian PSPTs. The intention was for the PSPTs to contact Helen ("meet the scientist") via e-mail and explore the variety of outreach programs offered to teachers based in schools in the vicinity of the US University. It was envisaged that Helen would suggest a number of hands-on activities to the PSPTs, of which they would select an activity and implement it during their 30-min PD presentation. The PSPTs did select one activity (from Turner et al., 2006), although it is the context of presenting this activity that is the focus of this chapter, particularly the larger collaboration that developed.

The Australian PSPTs, driven by their collective curiosities and self-imposed desires to implement a similar PD opportunity to that available to US teachers

---

[1] All names are pseudonyms.

teaching in the vicinity of the US University, negotiated a modification to their 30 min peer PD presentation assessment task. The modification was to remotely operate, in real time, an Atomic Force Microscope (AFM) that was located in the US University, in conjunction with the Helen and her colleagues. Although the assessment requirement was a 30 min PD presentation to 20 peers, these PSPTs extended the task to include the following five learning experiences: (a) participation in remote lectures from Helen and her colleagues, covering topics such as an introduction to nanotechnology and the Atomic Force Microscope; (b) participation in trials to operate the AFM remotely; (c) participation in a Faculty of Science 'Open day' presentation (to prospective students) at their Australian University, involving a hands-on activity relating to nanoscale measurement; (d) a visit to an Australian High School to deliver a lesson to Year 12 physics students (all the students in this class were boys, even though the school was co-educational) in conjunction with staff remotely from the US University; and (e) a presentation of their 'assessable' PD presentation in the form of a public seminar to an audience of over 100 people (instead of the intended audience of 20 peers). As an extension to the project, the PSPTs were invited to co-author a paper relating to the project. Two of the PSPTs took up this offer and the study was presented at a professional education conference.

Experiential Learning

This study is located within the theoretical framework of *experiential learning*. Experiential learning is a "dynamic view of learning based on a learning cycle driven by the resolution of the dual dialetics of action/reflection and experience/ abstraction" (Kolb & Kolb, 2012, p. 1215) and is said to be meaningful when there is personal involvement, self-initiation, and the freedom to explore (Houseal, Abd-El-Khalick, & Destefano, 2014). Experiential learning can frame a learners' immersion within disciplinary practice (such as authentic scientific practice) and takes the learner beyond substantive content. When learning is relating to a situation vicariously, such as the work of scientists, the learning is more often located within the related theoretical frame of inquiry-based learning (Chinn & Malhotra, 2002). In inquiry-based learning, the learning is a:

> ... student-centred, active learning approach focused on questioning, critical thinking and problem solving. Inquiry-based learning activities begin with a question, followed by investigating solutions, creating new knowledge as information is gathered and understood, discussing discoveries and experiences, and reflecting on new found knowledge (Savery, 2015, p. 11).

While there are many similarities between these two theories of learning, heavily based in action and experiences and reflecting on these actions and experiences. However, the use of questions to frame such learning is an important distinction between these two theories. In this instance, the idea was not to have the learners, in this case the group of PSPTs, involved in the process of doing scientific research with the US scientists on the AFM, but rather that they begin to develop an understanding of what the scientific process requires. In this sense experiential learning can connect the process of authentic scientific inquiry with inquiry-based learning. While the use of questions as a framing mechanism places a natural limitation on the scope of the project for the PSPTs, enabling them to explore questions they had developed themselves provides important ownership elements of this learning for the PSPTs. Experiential and inquiry learning, in partnership with a scientist, enables learners to participate in the processes of science: the PSPTs explored Helen and her colleagues' world in ways that enabled them to tell an authentic story about the scientists' work involving nanotechnology concepts for the purpose of learning science.

Engaging with scientists on aspects of scientific research provides a powerful context to engage teachers with scientific practices. Indeed, providing such direct experiences for pre-service teachers, which are modelled after ways we want them to teach their future students, has been highlighted as "an important trend" (NRC, 2007, p. 311). By extending the "meet the scientist" strategy into a partnership, the PSPTs and Helen created an experiential, authentic, inquiry-based learning situation that provided the PSPTs with access to the scientific community and enabled them to engage with scientific research processes. Such partnerships are often reported on in terms of the benefits to the students in the form of engagement with science, and the scientists in the form of additional resources for data collection efforts (Harnik & Ross, 2003; Wormstead, Becker, & Congalton, 2002). Further, Houseal et al. (2014) claim that the crucial engagement and intermediary role of the science teachers (or PSPTs in this case) often seems to either be taken for granted or subsumed by the assumptions underlying the partnerships. However, benefits to teachers include gains in content knowledge and an increased use of inquiry-based instructional strategies (see Caton, Brewer, & Brown, 2000; Evans, Abrams, Rock, & Spencer, 2001; Ross et al., 2003; Wormstead et al., 2002). In our study, we feature the PSPT as both a "student" learning about scientific research and learning about the art of teaching science, and as a "teacher" with the task of transforming what they learn into pedagogical performances that progressively approximate scientific practice.

Visibility of the PSPTs Learning

As already discussed in the vignette, the PSPTs' primary aim for their PD presentation was extended to include the remote operation of the AFM during a public seminar. Remote AFM operation was a regular occurrence between the US University and high schools in its immediate vicinity, and Helen and her colleagues were attracted by the possibility of remote operation of their AFM from Eastern Australia in real time—something never previously attempted. This chapter is interested in reporting on the emergence of the PSPTs' professional learning and identity in this context. As the US University was in a geographically different location and time zone (North Eastern USA), the PSPTs faced a number of challenges. We used the PSPTs' interviews conducted after the unit's completion and PSPTs' journal data to explore the challenges in their understandings of scientific practice for their pedagogical purposes, and were interested in what learning can eventuate from multi-setting work, and what it is to be a team member.

In terms of the present study, the PSPTs 'rewrote' the assessment task of a PD presentation to 20 of their peers. Instead, the PSPTs reset the activity to be a 'world first, real time' live operation of a US University's Atomic Force Microscope, presented as an evening public seminar. The PSPTs used their own intuitive skills to determine team member tasks and roles. Their individual skills complemented each other. For example, one of the PSPTs had the computer skills to negotiate firewalls and the download and installation of the software necessary for the remote operation of the AFM. Another PSPT initiated taking the project into the Australian High School setting. Yet another PSPT had the skills to adapt and modify a learning activity and present it equally well in both the Australian University setting and the Australian High School setting. The final two PSPTs held the group together, ensuring no individual deviated from the task. These two also conducted the background research. Thus, task division within the group evolved quite naturally, enabling the PSPTs to learn as individuals as well as to learn collectively as a team.

In terms of the present study, it is helpful to conceptualise the PSPTs' learning while at the Australian University, learning as a result of "meeting" Helen and her colleagues in their US-based University, and learning in an Australian high school as separate settings. As shown in Table 1, each setting was the site for multiple learning intentions. The PSPTs also felt the need to maintain their credibility with the Research Centre of the US University activity system (they did not want the withdrawal of the US commitment to their public seminar).

With respect to the PSPTs, each setting can be considered in terms of their learning and therefore each setting had a professional success outcome. Further, the three settings interacted with each other, and experiential learning was on offer in each.

Table 1 The nature of the PSPTs' intentions in the three different learning settings

Activity Setting	Intention 1	Intention 2	Intention 3
Australian University	To obtain at least a passing grade in the unit	To present a public seminar highlighting their successes	
US University	To determine if an "across the globe" outreach program was possible with pre-service teachers	To remotely attend and participate in the public seminar	To reflect on the experience of participating in a "world first", across the globe learning experience involving nanotechnology education
Australian High School	To expose year 12 physics students to a "real life" physics application	To attend a public seminar	

The PSPTs participated in discourses that questioned assumptions, expectations, and contexts to achieve deeper meanings and new perspectives that guided their actions. It was as a result of the multiple sites for learning that this rich range of learning opportunities was both enabled but can also be considered as of sites of conflicts and tensions due to contextual differences.

While there is evidence that the participants in both the Australian High School (Year 12 students and teacher) as well as the US-based scientists (Helen and her colleagues) benefited from the interactions with the Australian-based PSPTs, these learnings are not the focus of this chapter. Rather, the remainder of this chapter will explore the PSPTs' learnings generated by the interactions between the three settings, focusing on the PSPTs' pedagogical innovation and renewal.

The Learning Affordances of Social Processes

In this study we explored the effect of the social processes impacting on the PSPTs as they undertook a professional inquiry in 'real time' across the globe. The social nature of informal learning proved to be a critical aspect to the study. It included the interactions between the PSPTs and the US University-based scientists who initially provided ideas to the PSPTs for their PD presentation to their 20 peers. This subsequently developed into a collaboration between the PSPTs and the US scientists to present the PSPT-initiated "world-first" seminar. The study therefore explored how, after "meeting the scientist", the intention of 'doing a public seminar' enhanced the PSPTs professional learning.

> The journey has been very exciting and working with five other students was great, everyone got along well and we enjoyed the experience. Very different to past presentations—research what has been done before and replicating it. This was a first time, so it was hard, it was different and we didn't really know where it was going to take us. If I was just myself doing this, I never would have got through it! Especially the technology side of it. Everyone brought different strengths to the group to make it happen. (Jane: journal entry)

The initial contact with the US University-based scientists proved daunting for the group. Julie reflected in her journal that she felt "out of her league". She did not know how they would be received by the US scientists, and was concerned that she would be seen as dumb because she "didn't know what 'nano' was all about". Although all of the PSTs had the equivalent of an undergraduate science degree with a physics major, they were intimidated by the 'real scientists' and had a strong need to appear knowledgeable and capable. However, as a team, the group was strong enough to persevere and they overcame their fears.

Jane reflected in her journal:

> Had we not had Pete, I would have quit. We really needed him to sort out the technology side of things so we could show the scientists we were serious learners. I am no good with computers. I am scared I will deprogram them or something. But having seen Pete hack (well not 'real' hacking) into the computer to get us working, I saw a whole new side of him, and of computers. I still cannot do much with them, but they are not as scary.

In this reflection, Jane describes the difficulties the team had with the technology required for the remote operation activities. Remote operation of the AFM was commonplace between the US University and its local schools, but long distance remote operation, in real time, had not previously been tested. Additionally, the computer hardware and software were already established in the US University setting, but not in the Australian University setting or the Australian High School setting. Pete had to determine and establish the hardware and software in the Australian settings to enable the collaboration to continue. The profession learning of the PSPTs is explored below in terms of their philosophy of teaching, and professional relationships.

The philosophy of teaching of individual PSPTs became evident as they discussed aspects of the project. In his reflections and interview, Kieran showed depth and insight regarding his views on education. He said, "I have always been more interested in the big picture and seeing where things fit into life". This was quite evident in a reflection in his journal on the final Public Lecture, where Kieran wrote:

> I think the biggest issue for our group was determining the main focus of our project; nano-technology or using technology to bring experts like scientists and sophisticated/expensive equipment into the classroom. I fought with the majority of the group to make the focus of the presentation and the project the latter. I feel this could be revolutionary for science education and nanotechnology was just a topic we were using to demonstrate how technology could be used to bring an expert scientist in from anywhere on the planet to introduce or further the understanding of the topic.

Kieran wanted the Year 12 boys in the physics class, and his peers, to get more "real life" information out of the project. Kieran felt that education, in general, could be revolutionised by bringing experts into the classroom to show the "bigger picture" of how technology can impact on learning. Natalie, who commented in interview that she was now interested in, and had the confidence to try new things in the classroom, also held this view. She saw that teaching could bring things to life, and that she didn't have to base her instructional philosophy on her own past learning experiences. Natalie was also mindful of the pitfalls teachers could find themselves in:

> This opened my eyes to [how] in schools teachers can become blinkered to only what they teach and start to be unable to connect to other topics the students may be able to relate to with more ease.

These experiences indicate that both Kieran and Natalie did not just focus on the 'how-to' concerns of the classroom. Routine procedures like time management and lesson planning were present; however, the team learnt to reflect and work together to collaboratively build their professional teaching knowledge.

Professional relationships were explored and enjoyed by the PSPT group. There is no doubt the PSPTs enjoyed the move from being considered students, to being considered as "knowledgeable beings, with something to say that someone actually wants to listen to" (Natalie: interview). Julie noted that she was uneasy entering the Australian High School setting at first. She had had experience teaching physics in a school on her placement rounds, but didn't know if the "whole thing would work or not" (journal entry). However, once the team had overcome issues related to fire-walls and live link-ups, Julie found herself more at ease. The PSPTs also provided a morning tea for the science teachers in the Australian High School setting and this gave all involved an opportunity to discuss the project further. While Julie and Natalie initially felt uncomfortable mingling with the teachers, the morning tea and the fact that they had knowledge the teachers didn't have (but were curious about and wanted), enabled the break-down of barriers. Julie and Natalie both experienced life as professional teachers (albeit for a short time), and liked it.

Reflecting on professionalism as a teacher, Kieran reflected in his journal:

> The 5E teaching model tells us we must first engage the student by capturing the interests of students. How can we expect to get our students interested in something we cannot find fun in ourselves? This is why I thought it was important to include some humour, and I am glad my group supported and encouraged me to do so. I was happy when Steve [another US scientist] mentioned he believed it was important to be yourself and to give the students part of yourself. I can be professional with my students as well as be myself.

Two of the PSPTs experienced a change in identity and a feeling of professional-ism through writing a paper for a professional education conference. Funding was obtained from the Australian University setting for them to further explore their roles as researchers. They were both shy about the new level of professionalism offered to them through this extension activity: they were given office space and computers to conduct literature reviews, and were paid to explore their thinking. The following statement sums up the impact of this extended experience:

> I normally just read journal articles to get a bit of information for an assignment. You know, beef up the literature review a bit. Now I am not just using the articles, I am creating them. I find this scary to think that someone is interested in what we did, what we learned. Yea, it was a world first, so I guess it is interesting. But me? Doing something and writing about it? Scary, but good. I like the importance feeling. (AngusL journal)

In other words, the activity of "meeting the scientist" led to two undergraduate PSTSs being able to experience academic life. The other group members were given the same opportunity but declined for personal reasons. Overall, each of the PSPTs had something to say, as do all pre-service teachers. They just needed support to gain the confidence to acknowledge the contribution to teaching and learning that they themselves had made.

Learning from the Obstacles

The PSPTs had to find ways to 'work around' obstacles by adapting their practices. Natalie reflected in journal on the obstacles in general:

> Most of the obstacles we came across we could learn from. All the things that made our presentation harder were also things that could come up when planning any lesson on something completely new. The fact that we were doing something that had never been done before, exploring something none of us had ever heard of, in a way that included computers, made this a very daunting task!! However the fact that I came out alive and with

new knowledge has probably given me the confidence to try new things and take on challenging opportunities when I get into schools. I'm glad we took the challenge of sending that email to Helen [US scientist]. The result has been phenomenal.

What was initially a small topic for six PSPTs, very quickly became a "world first" experience. Although it proved to be a difficult task with many obstacles and tensions to deal with, Jane does not regret it—she reported in her journal being "Glad for the experience and hope everyone gets a chance to do something like this". The six PSPTs all experienced high levels of personal and academic growth, as summarised by Julie:

> It was part of our progression from having almost no knowledge, to having enough knowledge and confidence to present in a school and then critiquing our performance in the classroom and making changes before our final public seminar. We all learnt a lot … Importantly because of going into the school with almost no guidelines and having to present something credible we learnt extensively about teaching and co teaching (interview).

Final Comments

Our focus within this chapter is on aspects of the PSPTs' learning. The six PSPTs were accessing three learning sites simultaneously—one in North Eastern USA, and two in Australia. The PSPTs developed strategies to allow them to resolve difficulties, overcome obstacles and experience professional learning. They gained new confidence in their teaching and planning abilities; they were able to resolve contextual differences as they could draw on each other and a supporting group of scientists on the opposite side of the globe; and some took up the continuing challenge to be assisted in developing identities as researchers of their own teaching and learning. While the learnings of the PSPTs are enmeshed with the learnings of the Australian High School teacher and Year 12 students, and the US University-based scientists, this complexity is beyond the scope of the current work.

The findings as discussed here have important implications for teacher education. Specifically, we feel that deliberate planning for networking between pre-service science teachers and scientists may well assist in the development of professionalism and teaching philosophies of pre-service science teachers. Proximity of settings need not be a hindering issue, as we have demonstrated. By undertaking this study, we have also discovered that it is perhaps more important for the pre-service science teacher to develop the ability to engage in ill-defined problems with scientists than to be concerned with how much they know or whether they have acquired skills. It therefore seems fitting to leave the last words with a participating PSPT:

> Our group came so far, and overcame so many obstacles that if I end up being placed in a similar situation in a school (being told to teach something I know nothing about), I will not resign, I will know where to start looking for help. What is more, I think we changed the minds of many of our classmates. Their changed views were best summed up by a statement on one feedback sheet: "Whoever knew physics could be so much fun … maybe I should give Biology away for Physics!" (Julie: interview)

Acknowledgements We are thankful for the most generous support from the US University setting and its associated Research Centre. Without their after-hours support (due to time zone differences) the professional learning of the PSPTs would not have been possible. We further acknowledge the PSPTs for their stoic endurance throughout the semester and beyond. Without your individual perseverance, the collective could not have succeeded. Finally, we would like to acknowledge the Australian University's CLI Cluster Grant for funding to allow two PSTS to "tell their story".

References

Caton, E., Brewer, C., & Brown, F. (2000). Building teacher-scientist partnerships: Teaching about energy through inquiry. *School Science and Mathematics, 100*(1), 7–15.

Chinn, C. A., & Malhotra, B. A. (2002). Epistemologically authentic reasoning in schools: A theoretical framework for evaluating inquiry tasks. *Science Education, 86*, 175–218.

Evans, C. A., Abrams, E. D., Rock, B. N., & Spencer, S. L. (2001). Student/scientist partnerships: A teachers' guide to evaluating the critical components. *American Biology Teacher, 63*(5), 318–323.

Harnik, P. G., & Ross, R. M. (2003). Developing effective K-16 geoscience research partnerships. *Journal of Geoscience Education, 51*(1), 5–8.

Houseal, A. K., Abd-El-Khalick, F., & Destefano, L. (2014). Impact of a student–teacher–scientist partnership on students' and teachers' content knowledge, attitudes toward science, and pedagogical practices. *Journal of Research in Science Teaching, 51*(1), 84–115.

Kolb, A. Y., & Kolb, D. A. (2012). Experiential learning theory. In N. M. Seel (Ed.), *Encyclopedia of the science of learning* (pp. 1215–1219). Dordrecht, The Netherlands: Springer.

NRC. (2007). *Taking science to school: Learning and teaching science in grades K–8.* Washington, DC: National Academies.

Ross, R. M., Harnik, P. G., Allmon, W. D., Sherpa, J. M., Goldman, A. M., Nester, P. L., & Chiment, J. J. (2003). The mastodon matrix project: An experiment with large-scale public collaboration in paleontological research. *Journal of Geoscience Education, 51*(1), 39–47.

Savery, J. R. (2015). Overview of problem-based learning: Definitions and distincitions. In A. E. Walker, H. Leary, C. Hmelo-Silver, & P. A. Ertmer (Eds.), *An essential reaing in problem-based learning* (pp. 5–16). West Lafayette, IN: Purdue University Press.

Tsui, A. B. M., & Law, D. Y. K. (2007). Learning as boundary-crossing in school-university partnership. *Teaching and Teacher Education, 23*(8), 1289–1301.

Turner, K., Tevaarwerk, E., Unterman, N., Grdinic, M., Campbell, J., Chandrasekhar, V., & Chang, R. P. H. (2006). Seeing the unseen: The scanning probe microscope and nanoscale measurement. *The Science Teacher, 73*(9), 58–61.

Wormstead, S. J., Becker, M. L., & Congalton, R. G. (2002). Tools for successful student–teacher–scientist partnerships. *Journal of Science Education and Technology, 11*(3), 277–284.

Trial-and-Error, Googling and Talk: Engineering Students Taking Initiative Out of Class

Elaine Khoo and Bronwen Cowie

Abstract This chapter reports on the strategies that first year engineering students used to supplement and extend their laboratory and lecture learning about a 3-dimensional computer-aided design (3D CAD) software, *SolidWorks*. A capacity for self-initiated and self-directed learning as part of developing lifelong learning capabilities is widely recognized as a critical outcome for today's engineering graduates (Jamieson & Lohmann, 2009; National Academy of Engineering, 2004). This capacity naturally spans both formal and informal settings. We illustrate what this might look like drawing on two projects. One investigated the role ICTs/e-learning can play in tertiary teaching and learning (Johnson, Cowie, & Khoo, 2011) and the other investigated the nature, development and implications of software literacy (Khoo, Hight, Torrens, & Cowie, 2016). Engineering students in these studies reported a diverse array of self-initiated and self-directed informal learning actions including daily conversations with peers, out-of-class conversations with lecturers, trial and error in their own time, work through course materials, and use of YouTube videos and dedicated online professional discussion forums. Different students expressed a preference for different combinations of these approaches. Student informal learning therefore covered a patchwork of learning processes and outcomes within their formal learning programmes. Students asserted informal learning activities were essential to enrich and complement formal learning occasions if they were to develop adequate/sufficient understanding of and competency in the use of software to solve engineering design problems.

In the chapter we pay particular attention to what students have to say about why they initiate these informal leaning activities and the significance students place on these activities. We conclude the chapter by speculating on implications

E. Khoo (✉) · B. Cowie
University of Waikato, Hamilton, New Zealand
e-mail: ekhoo@waikato.ac.nz; bronwen.cowie@waikato.ac.nz

© Springer International Publishing AG, part of Springer Nature 2018
D. Corrigan et al. (eds.), *Navigating the Changing Landscape of Formal and Informal Science Learning Opportunities*,
https://doi.org/10.1007/978-3-319-89761-5_12

for teaching and learning of blurring the formal-informal boundary, in particular the contribution informal learning has to make to learning in contexts that are usually seen as "formal".

Keywords Engineering learning · Higher education · Informal learning

Introduction

An assumption underpinning all of the chapters in this book is that informal learning processes are a fundamental aspect of many everyday activities. However, their relevance and contribution in more formal learning and educational settings has not always been recognised. Increased interest in informal learning can be linked with the impact of increases in the power and ubiquity of information and communication technologies (ICTs). These have increased and diversified the means individuals and groups have to access sources of information and support for learning. These offer formal learning agencies and processes the option of purposively leveraging resources such as Facebook, Twitter and the Internet alongside and in addition to providing formal supports for learning.

Increased interest in informal learning can also be linked with the current political imperative to foster student learning capacities and inclinations as part of preparing individuals who will prosper in the "knowledge society" (e.g., Bell, Shouse, Lewenstein, & Feder, 2009). The European Union position paper *The future of learning: preparing for change* puts forward a vision of learning as a lifelong, life-wide process and notes that, "The overall vision that accompanies this [conceptualisation of learning] is that personalisation, collaboration and informalisation (informal learning) will be at the core of learning" (Redecker et al., 2011, p. 9). Building on this vision, Jackson (2014) asserts that "one of the most important things higher education can do to prepare adult learners for learning in the rest of their lives is to pay greater attention to the informal dimension of their learning lives while they are involved in formal study in higher education" (p. 1). Informal learning need not, and should not, be something that occurs after/outside of formal learning. Informal learning can usefully take place alongside and in combination with more formal learning activities as evidenced by current scholarship that calls for broader conceptualisations of learning (e.g., Barron, 2006; Stocklmayer, Rennie, & Gilbert, 2010; and this book).

In the chapter we pay particular attention to what tertiary engineering students have to say about why they initiate informal learning activities to complement formal learning activities and the significance they place on these activities. We are particularly interested in the network of resources and supports that learners can—and need—to develop to assist them to learn, and keep on learning, across the various contexts of their lives (Dierking, 2015). Specifically, we focus on the learning actions and strategies employed by engineering students outside of their scheduled formal learning activities (their lectures and labs) in learning how to use SolidWorks, a CAD software package. We conclude the chapter by speculating on the implications for teaching and learning when the formal-informal boundary blurs, and the nature and role of the "learning ecologies" (see chapter "Viewing Science Learning Through an Ecosystem Lens: A Story in Two Parts" by John Falk and Lynn Dierking, also Barron, 2004) students develop in support of their own learning and personal development.

Establishing a Framework

Formal learning, by design, is where learners engage with ideas and materials developed by a teacher as part of a programme of instruction. In tertiary settings formal learning tends to be associated with structured and didactic, teacher-led pedagogies aimed towards a particular end goal (Willems & Bateman, 2013). Informal learning, on the other hand, is usually understood to be unstructured, self-directed, emergent and linked with an individual's work-related, family or leisure activities (Dierking, 2015; Halliday-Wynes & Beddie, 2009). Typically, informal learning involves a combination of information seeking, observing, help seeking, asking questions, trial-and-error and so on (Siemens, 2004). On the whole the initiative for informal learning starts with learners as they seek to deepen and extend their learning and understanding (Jackson, 2013); it is the learner who takes responsibility for and ownership of the learning and its progress (Falk & Dierking, 2010; Marques et al., 2013). This is congruent with understandings that in order to develop deep competences, learners must be motivated to do so otherwise they may simply cover content as a means to fulfill formal assessment criteria (Marques et al., 2013). Informal learning then happens in accordance with what is known variously as intrinsic motivation to learn (Boekaerts & Minnaert, 1999), self-regulated learning (Zimmerman, 2000), adaptive help seeking (Karabenick, 2003), and self-directed learning (Gillet, Law, & Chatterjee, 2010). The question this chapter addresses is how students blur the line between formal and informal learning as part of self-directed learning. (Readers are also referred to chapter "Using a Digital Platform to Mediate Intentional and Incidental Science Learning", by Cathy Buntting, Alister Jones and Bronwen Cowie, for a discussion of incidental and intentional learning.)

A Learning Ecologies Approach to Self-directed Learning

The rise of the Internet has made a significant impact on students' capacity for self-directed or free-choice learning by making it easier for individuals to find and access "resources and activities that can support their learning on their own terms" (Barron, 2006, p. 194). This has contributed to a rethinking of some of the assumptions about informal learning. In particular, attention has turned from an emphasis on the physical context of learning to the resources and strategies that individuals and groups can marshal to support their interest-driven learning (Falk & Dierking, 2010). The notion of a "personal learning ecology" is one way of making sense of the strategies and resources people use to progress their learning across time and settings.

In this chapter, we define a learning ecology as encompassing the contexts, relationships, strategies and resources that an individual mobilises to achieve a personal learning goal. The notion of a learning ecology takes into account that the boundaries between contexts tend to be permeable, and that people draw on multiple relational and material resources to meet their current needs, no matter where they happen to be. Each and every context offers a "unique configuration of activities, material resources, relationships, and the interactions that emerge from them" (Barron, 2006, p. 195). Both physical and virtual contexts can provide opportunities and supports for self-directed learning and the 2014 *Horizon Report for Higher Education Preview* (New Media Consortium, 2014) emphasises that social media tools can provide a useful, and in some cases preferred (Moll, Nielsen, & Linder, 2015), way of accessing support from a networks of peers/friends and experts.

Barron (2006) highlight that more experienced students access and use a wider range and types of learning strategies and resources even when their access to physical and virtual resources is the same. They suggest differences might be due to variations in learner knowledge and interests, learner perceptions of the interdependencies between different resources, and learner resourcefulness (their capacity, inclination and persistence in identifying resources). Staron (2011) notes that in order to establish a learning ecology that is meaningful, authentic and supportive of their growth and personal well-being, learners have to have the courage to do what is most appropriate and useful in establishing and activating a network of supports. Learners therefore need to have the confidence, courage and capability to identify and pursue strategies that will support their learning and learning progress—and knowing how to create and sustain a learning ecology is an essential part of "knowing how to learn" in all the different contexts that comprise an individual's life (Jackson, 2013, p.1).

A Holistic Perspective for Engineering Education

The Washington Accord (2013) is an international agreement among professional engineering institutions that confer accredited qualifications in professional engineering. It details the broad range of graduate attributes and professional competencies that today's engineering graduates need. Specifically, it states that the fundamental purpose of engineering education is to build each graduate's knowledge base and attributes so they can continue learning and develop the competencies required for independent practice beyond formal learning contexts. In this it recognises that graduates need to develop the capacity for self-directed lifelong learning for them to function effectively in an ever changing and increasingly complex world. In considering how to achieve these goals, Scott and Yates (2002) remark that it is important to "focus on the entire undergraduate experience rather than just what is taught" (p. 363). With this in mind, we examined the potential of an ecological approach for understanding engineering student experiences of learning CAD software across both formal and informal contexts.

A Case Study from Engineering Education

In this chapter we draw on data from a Government-funded project *Copy, cut and paste (CCP): how does this shape what we know?* (Khoo, Hight, Torrens, & Cowie, 2016) to report on the views of participating tertiary engineering students from the University of Waikato, New Zealand. The CCP study aimed to explore the development of software literacy in two tertiary teaching-learning contexts: mechanical engineering and media studies. We defined *software literacy* as involving expertise in using and critiquing the influence of discipline specific software in pursuit of particular learning and professional goals. Our premise was that developing the ability to problem solve and critique software is an essential proficiency in our software-saturated culture. For the purposes of this chapter, we report on the findings from the mechanical engineering case study. We were particularly interested in what enabled and constrained engineering students' learning to use a computer-aided design (CAD) software, SolidWorks. The use of CAD software is accepted practice in modern engineering and SolidWorks is used extensively in engineering industries across Australasia.

No entry-level familiarity with CAD or 3D drawing software is assumed at the onset of the degree programme, although students are expected to be familiar with the use of computers. In year 1 the students receive around 6 h of formal instruction on SolidWorks. In year 2 students can choose to attend a series of three-hour supervised computer laboratory sessions where they complete tasks to develop their proficiency with SolidWorks. The course lecturer and tutors are available to assist students with any issues, difficulties and questions. Students can also use the computer labs at the university in their own time, and can opt to install SolidWorks on their personal computers and work through the same tasks at home. Each of the assigned tasks is assessed. As part of the year 2 course students are also required to collaborate on a group design project, which they present at a Faculty open day. The aim of the project is for students to develop and demonstrate their SolidWorks-supported design understanding and application.

Between their second and third year students spend 10 weeks on work placement in an engineering firm where they may be required to use SolidWorks or other similar CAD software. Third year course assessments require students to use SolidWorks in order to develop and build an artefact, such as a conveyor belt system, that incorporates the engineering design principles they have learned. These artefacts are exhibited and judged as part of a Faculty open day. In year 4 students are expected to use SolidWorks for individual projects.

Data were collected through observations of the year 2 students during the SolidWorks labs and interviews with the lecturer and tutors. We also surveyed students about the ways they learned SolidWorks in and out of labs and conducted a focus group with six volunteer students to elaborate on the survey results. Additionally, we conducted a focus group with seven volunteer year 3 students on if and how they had been able to use what they had learned about SolidWorks during their work placement. We wanted to see if and how students drew from similar strategies to those they had reported the year before. These focus groups took place soon after the students' work placement. We then observed the year 3 students working with SolidWorks as part of their coursework and conducted a focus group with seven students at the end of the year. A separate group of six elite year 4 students was interviewed. Each year our university is represented by a team of fourth and final year students in a prestigious international Formula SAE-A competition highly regarded by the industry. Each team must design, build and race a small high-performance race car. Those in the team are considered to have developed sophisticated software literacy skills.

Table 1 illustrates the data collection focus across the 3 years of study (Year 2–4). For this chapter, we are not presenting data from the year 1 students because the research focus for them was not on the development of their learning ecology.

Below, we report on the strategies and resources the Years 2–4 students described as supporting their use of and problem solving with SolidWorks.

Table 1 Description of the formal learning opportunities and data collection across the three years of study

Year level	Formal focus of solidworks learning/ use	Data collection procedure	When collected
Year 1	Introduction to SolidWorks including lab-based learning and structured group project work	No data were collected	
Year 2	Lab-based learning followed by structured group project work to extend students' use of SolidWorks in engineering design	Student survey, student focus group, lecturer interview, tutor interview	At the end of the course
Year 3	Work placement—on the job use of CAD	Student focus group	Immediately after the completion of work placement
Year 3	Advanced individual lab-based structured exercises and a real-world group project using SolidWorks	Student focus group, lecturer interview	At the end of the design course
Year 4	Integrated use (all students)		
	Elite group of students design a racing car for an international competition	Student focus group comprising elite student group	During the design work

Changes in Student Learning Preferences and Ecologies

Student commentary over the years of study indicated an increase in the sophistication of the kinds of pedagogical, technological and learning resources they accessed to productively learn to use and problem solve with SolidWorks. While the formal learning place was the university, the students' learning ecology encompassed a variety of material and relational resources and strategies. We begin by outlining the strategies the Year 2 students used.

Year 2 Students' Learning Preferences and Ecologies
Year 2 student survey data indicated that they used a variety strategies when learning to use SolidWorks (see Fig. 1).

Almost 40% of students said their preferred problem solving approach would be to 'ask the lecturer for help'; for another 22% their first preference would be to 'go the Internet' or 'check their lab notes' for specific help. 'Lab notes' and/or the 'SolidWorks (online) manual' was the second choice for 37% of the students; nearly a quarter (24%) would 'ask a friend/peer' as their second choice. The strategies of 'watching someone using SolidWorks', 'trial-and-error' and 'Internet video tutorials' were the third most preferred choice for about a sixth of all participants.

The responses suggest that these Year 2 students tend to draw first on "official" and formally recognised authoritative sources of assistance. The lecturer and materials developed by the lecturer and/or support materials that were part of the software package were privileged. 'Asking peers' and 'trial-and-error' featured less. Alongside, and somewhat in contradiction of these responses, three quarters (76%)

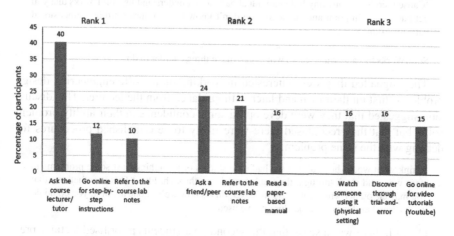

Fig. 1 Rankings of strategies for learning SolidWorks (n = 67)
Note: Rank 1 to Rank 3 denotes, in descending order, the usefulness of various strategies for learning SolidWorks as reported by students (Taken from Khoo, Hight, Torrens, & Duke, 2014)

of the survey respondents reported installing SolidWorks on their own computer so they could use and practice using SolidWorks for their coursework in their own time. Just over a quarter (27%) reported using SolidWorks outside of formal coursework for a range of recreational purposes. An example (from the open ended survey response) was:

> I have many sketches which I have a hard time imagining in 3D therefore I use SolidWorks to give me a more detailed version of what I have imagined.

In focus groups, students reiterated they drew on a variety of resources and strategies to help them learn to use SolidWorks when working in a self-directed way. These included working through the tutorials embedded in the software, drawing on 'more expert' peers, discovering through trial-and-error, as well as using online materials such as YouTube instructional videos. A majority of students (84%) reported they were comfortable in engaging with new technologies, hence their willingness to explore more informal ICT-supported forms of resources. These notably involved developing expertise in finding and identifying instructional material suited to 'their level'.

Students actively drew on help from more expert peers within the formal setting as exemplified in this comment:

> I've been working next to a fourth year I'm friends with and he's looked at my work and gone, "Whoa, dude, hold on—let me show you how to do this" and he's stepped in and shown me a whole bunch of stuff.

The focus group students also recognised that learning to use SolidWorks required an investment of time for learning, stating:

> 'Cause there's so many tiny little individual parts about understanding SolidWorks that you get past a certain point and suddenly you don't know how to mirror a three-dimensional part, for example.

> SolidWorks has a learning curve which can make things harder to do.

They signalled they were interested in working on real-life engineering design problems, first on their own and later with guidance from the lecturer. This discussion suggested that they were developing some confidence in their ability problem solve but that they recognised there were likely to be established procedures for dealing with the same problem:

> I think what would be cool is if we had case studies or something; just some problems in class we could work through, the teacher could go through, like, "This is something that you may encounter while you're doing CAD, this is how we've gone about it, you could do it your way but this is the procedure we've used …

Overall, there was a sense that the second year students prioritised lecturer prepared and/or authorised resources although they also sought out help from their peers and accessed online materials. Their survey data and comments indicated that they appreciated the need and value of a diversity of supports as part of a learning ecology focused on becoming more proficient in learning SolidWorks.

Year 3 Students' Post Work-Placement and End of Year Reflections
As a learning ecology has a contextual aspect we probed students' experiences of using SolidWorks while on work placement. We were interested in if and how they drew on the resources and strategies they had described to us as year 2 students. The year 3 focus group highlighted the range of strategies and people that they had to draw on as part of their learning ecology. These included an instructor-prepared reference sheet on key SolidWorks operations, seeking out help online, and asking workplace peers/colleagues:

> I guess in the labs [at university] you could get help from the demonstrators but it's 8 am and they were taking a long time to come so I didn't, you know. Yeah, so I felt like the flip sheet [reference sheet], where it's there when you want it; like, going online is good but you waste a lot of time going online as well so that's why I tended to ask people at work. When I was doing the course at home I was always on the [SolidWorks help] forums and just general how-to's on the Internet.

This student further discriminated between strategies that were helpful when he was a novice and strategies after he had developed more independent and advanced troubleshooting skills.

> If you were just beginning SolidWorks then the [online] tutorials that come with it [the software] is a good place to start. But when you get a bit more in-depth it sort of loses its value. Yeah. Just asking people, especially if they know what they're doing, is the best way, I've found—that one-on-one sort of tuition.

After their placements, the students commented that it was not adequate to depend solely on university coursework to understand the various aspects and potential of the software in a workplace. They spoke of the value of practical and context-specific one-to-one assistance from more experienced industry experts when thrown into challenging real world contexts:

> On my first day, I think, I was sat down and he was like, "Right, make this" and I made it and he was like, "That's totally wrong" and then spent like three days teaching me how to use it, just how he liked it.

Another student elaborated:

> In my work placement I had a couple of people who knew how to do everything so I would ask them ... there was some stuff that they didn't know and there were some things that I'd learnt at Uni that they didn't know existed in SolidWorks so it's kind of interesting when you see people's overlap because they were self-taught as well. Yeah, they just always seemed to show me how to do it a lot easier than what I was doing it.

Threaded throughout these three comments is the idea that there are more and less efficient ways to get the job done. Subsequent group discussion indicated students were well aware of the need for persistence when working through a problem with SolidWorks:

> I would say [I am] competent [in using SolidWorks] but I can be easily tripped up, and get stuck. I guess when I encounter a problem it does take me quite a while to get around it. If it's really pear-shaped or screwed up, you've got to sit there and nut it out.

The second focus group interview with the 3 year students occurred towards the end of the academic year. We were interested to see if there was any change in the ways students conceptualised and used a personal learning ecology to support advanced software use. Student commentary indicated that at this time students generally started troubleshooting by referring to Internet resources such as YouTube. They felt confident to do this because they had a knowledge base to draw from:

> We had the base knowledge and it was generally pretty easy if you needed a little extra help. We had enough, like, knowledge to follow a tutorial [on YouTube] pretty easily.

Again, persistence to work through a challenge was seen as important, as raised by another student:

> Yeah, if you do strike a problem generally you can just muscle through it, it may take a bit longer.

One student reflected on the strategies he had developed when highlighting the value of persistence and troubleshooting when working through advanced coursework:

> From [first and second year] we pick up all the basic stuff and learn how to do it, but during that process we learn how to use the troubleshooting method and that's I think the most valuable thing that helped me later on … I'm confident with even something I don't know, I know how to find it, how to learn it from online resources, then I can still make that happen [on SolidWorks]. I think that's the most valuable thing, that even later when I go to my fourth year and do some more complicated thing, I know where to go, I [won't be stuck] and waiting for someone to help me. I can still go through my work and it may take a little bit of a long time but at the end of the day I'll probably still pick it up.

The themes emerging from the year 3 interviews affirmed that students used and valued a variety of informal learning resources and strategies. Students accessed and used Internet resources, peers, and dogged persistence as part of their expanding learning ecology. A developing confidence in their own ability to troubleshoot underpinned the interplay of the learning resources and the strategies they drew from in order to be able to use SolidWorks to solve the more challenging problems associated with its use in real-world contexts. As they became more competent they relied more on their own resources and capacity to learn, and less on lecturer-prepared and commercial materials.

Elite Year 4 Students' Learning Preferences and Ecologies
The elite year 4 student group provided insights into the developmental trajectories that had contributed to their proficiency with SolidWorks. The trajectories they described involved an increase in understanding of the efficiency and sophistication of the design features that SolidWorks could be used to accomplish. They emphasised that "you've got to learn the foundations to do it effectively". One student explained:

> Once someone teaches you the basics of sketches and you learn those things then you can start experimenting and troubleshooting and stuff and then using the different features and that gets you nice and efficient.

Another student elaborated, "There's often several ways of doing something and it's learning the most efficient way". Again, discussion focused on more than the need to produce a functional solution—the aim was an efficient design.

Students indicated that they continued to refer to the SolidWorks online manual, noting that it was comprehensive but not their first choice as an information source. The SolidWorks in-built tutorials were described as "good" for scoping out ideas although they were easy to get "lost" in. One student explained:

> The [SolidWorks] tutorials are good for getting ideas when you start modelling [3D components].

The group endorsed the value of trial-and-error as a problem solving strategy:

> Students have definitely got to muck around through trial and error. They would really struggle if they just went in, did the stuff and then just went home. ... You just muck around and change some things and it works. So next time you go, "Well I did this last time and it worked".

Students used an Internet search (Google) to find answers to specific questions such as "Why won't my surfaces merge?", commenting that as Google searches generated a "million reasons" they needed to decide "do any of these apply to you?" They were clear that it was essential to critically engage with the various sources of help and support that were available.

Student representation of the learning process as a multifaceted activity was best reflected in the following extended comment by one of the students:

> Probably one of the big things that's kind of cool is that we can take it [SolidWorks] home, use it on our personal computers at home, and come into the lab. You can work on assignments at home and play around on it at home. Another way that I picked up [ideas] was from working around people... working together and knowing how to do things better. You don't get that sort of support from the teachers. I guess if you go up and ask them, they probably will give you a hand, but most of my learning on SolidWorks has been done by working on it at home or playing around at home and learning from peers and also YouTube videos. If there's no one around and you can't do it, type it into Google, type it into YouTube, and hopefully you'll get something and if you don't then ask for some help.

A clear theme emerging from the year 4 focus group of students was that students had to want to "learn to drive the programme" and that they needed to invest personal time, beyond class time, to achieve this. Overall their view was, "You've got to be doing it independently as well, like the other guys have said. It's not something you can just pick up just from the class". Here we can see the extent to which this group of elite students was engaging in self-directed learning. The consensus was that over time they had taught themselves by doing tutorials, experimenting with the programme and watching how other people, including their tutors, went about completing design tasks. Their view appeared to be that in the long term they, and engineers, should be able to "teach" themselves:

> You get to a level where you're capable—and I suppose that's just engineers as well—you're capable to teach yourself. You can use different resources just to teach yourself.

The significance of this understanding was confirmed by the lecturer who taught year 2 and 3 engineering design. He emphasised that students need to learn how to learn to problem solve. His expectation was that students would build on the more formal university-provided instruction and take up opportunities to learn SolidWorks through informal means.

Discussion and Conclusion

Current engineering professional standards emphasise the need for engineering graduates to develop capacity for self-initiated and self-directed learning (Jamieson & Lohmann, 2009; National Academy of Engineering, 2004; Washington Accord, 2013). In this chapter, we have scoped the kinds of learning strategies and resources that engineering students used to supplement and extend their laboratory and lecture learning about using SolidWorks, a CAD tool. Students were clear that classroom time and materials designed to supplement the formal curriculum were useful but insufficient. The students we spoke with also emphasised the need for persistence in the face of challenge. They identified this and their own ability to be self-directed when working towards a solution as elements of their learning and learning to learn.

Early in their engineering programme students placed considerable value on formal supports but as time went by they made more use of help from peers, online resources (SolidWorks tutorials, Google, YouTube), expert others and trial-and-error. The fourth year students indicated they had developed a sophisticated learning ecology that included the capacity to critically gauge which resources were appropriate to their learning level and the task design aims. In contexts comprising the university, workplace and home, our participants described learning ecologies consisting of multiple relational and material resources—knowledgeable peers and workplace colleagues and Internet-based resources of various kinds.

For us, students' drawing on help from peers and online resources across formal and informal settings raises questions about the distinction between the two. Although these findings focus on engineering students' learning of a disciplinary specific software package, some key ideas can be distilled as implications for other disciplines and tertiary institutions as a whole. For example, we agree with Jackson (2013) and Redecker et al., (2011) that a learner's ability to create their own ecology for learning and development is a crucial capability in today's complex and dynamic world. This means lecturers have an obligation to help students become aware of the need to develop a repertoire of learning strategies and resources.

Concomitantly, they have an obligation to provide students with opportunities to develop this repertoire. As part of this, they need to foster students' confidence, courage and resilience to learn—and to learn how to learn—new ideas. These are an essential element of a learning ecology (Staron, 2011). For many lecturers in tertiary education fostering student learning ecologies will involve a shift from a lecturer as a dispenser of knowledge to someone who more proactively supports students to take on more independent and critical roles in their own learning—and this has implications beyond individual lecturer change.

Institutions need to consider how they might assist students to develop productive lifelong learning capacities for the twenty-first century work and leisure environment. We wonder if, how and to what effect the notion of a learning ecology might be incorporated into institutional graduate profiles as a tool for assisting lecturers and students to develop the network of learning strategies and supports (both formal and informal) essential to individuals being able to learn lifelong and lifewide. As tertiary institutions increasingly move to exploit the teaching and learning potential of e-learning and social media platforms, we need to be aware that new and different kinds of learning and learning ecologies will become possible. This offers an exciting space for further research, policy and practice development.

Acknowledgements We gratefully acknowledge funding support from the Teaching and Learning Research Initiative, Ministry of Education, New Zealand.

References

Barron, B. (2004). Learning ecologies for technological fluency: Gender and experience differences. *Journal of Educational Computing Research, 31*(1), 1–36.

Barron, B. (2006). Interest and self-sustained learning as catalysts of development: A learning ecology perspective. *Human Development, 49*, 193–224.

Bell, P., Shouse, A., Lewenstein, B., & Feder, M. (2009). *Learning science in places and pursuits.* Washington, DC: National Research Archives.

Boekaerts, M., & Minnaert, A. (1999). Self-regulation with respect to informal learning. *International Journal of Educational Research, 31*(6), 533–544.

Dierking, L. D. (2015). Learning science in informal contexts. In R. Gunstone (Ed.), *Encyclopedia of science education* (pp. 607–615). Dordrecht, The Netherlands: Springer.

Falk, J. H., & Dierking, L. D. (2010). The 95% solution: School is not where most Americans learn most of their science. *American Scientist, 98*, 486–493.

Gillet, D., Law, E. L. C., & Chatterjee, A. (2010). Personal learning environments in a global higher engineering education Web 2.0 realm. In: *1st IEEE Engineering Education Conference (EDUCON).* Madrid, Spain. Retrieved from http://www.role-project.eu/wp-content/uploads-role/2010/01/educon_dg_final_free.pdf

Halliday-Wynes, S., & Beddie, F. (2009). *Informal learning. At a glance.* National Centre for Vocational Education Research Australia. Retrieved from http://files.eric.ed.gov/fulltext/ED507131.pdf

Jackson, N. (2014). *Ecology of lifewide learning & personal development.* Keynote presentation at the University of Brighton's Annual Learning and Teaching Conference. Retrieved from http://about.brighton.ac.uk/clt/files/3014/0422/8832/Ecology_of_Learning_and_Development_Handout.pdf

Jackson, N. J. (2013). The concept of learning ecologies. In N. J. Jackson & B. C. Cooper (Eds.), *Lifewide learning, education and personal development.* E-book available from www.lifewideebook.co.uk

Jamieson, L. H., & Lohmann, J. R. (2009). Creating a culture for scholarly and systematic innovation in engineering education: Ensuring US engineering has the right people with the right talent for a global society. *American Society of Engineering Educators (ASEE), 30*(17), 246–251.

Johnson, E. M., Cowie, B., & Khoo, E. (2011). *Exploring elearning practices across the disciplines in a university environment* (Summary Report). Wellington: Teaching Learning Research Initiative. Retrieved from http://tlri.org.nz/exploring-e-learning-practices-across-disciplines-university-environment

Karabenick, S. A. (2003). Seeking help in large college classes: A person-centered approach. *Contemporary Educational Psychology, 28*(1), 37–58.

Khoo, E., Hight, C., Torrens, R., & Cowie, B. (2016). *Copy, cut and paste: How does this shape what we know?* Final report. Wellington, New Zealand: Teaching and Learning Research Initiative. Available at: http://www.tlri.org.nz/tlri-research/research-completed/post-school-sector/copy-cut-and-paste-how-does-shape-what-we-know

Khoo, E., Hight, C., Torrens, R., & Duke, M. (2014). "It runs slow and crashes often": Exploring engineering students' software literacy of a computer-aided design software. In A. Bainbridge-Smith, Z. T. Qi, & G. S. Gupta (Eds.), *Proceedings of the 25th Annual conference of the Australasian Association for Engineering Education (AAEE2014)*. Palmerston North, New Zealand: School of Engineering & Advanced Technology, Massey University.

Marques, M., Viegas, M., Alves, G., Zangrando, V., Galanis, N., Brouns, F., Waszkiewicz, E., & García-Peñalvo, F. (2013, November 06–08). Managing informal learning in higher education contexts: the learners' perspective. In *ICBL2013: International Conference on Interactive Computer-Aided Blended Learning*. Florianópolis, Brazil. Retrieved from http://dspace.learningnetworks.org/bitstream/1820/5150/1/ICBL_2013_final_78.pdf

Moll, R., Nielsen, W., & Linder, C. (2015). *Physics students' social media learning behaviours and connectedness*. Paper presented at NARST, Chicago. Retrieved from http://wordpress.viu.ca/mollr/files/2015/04/NARST-2015-paper-submitted.pdf

National Academy of Engineering. (2004). *The engineer of 2020. Visions of engineering in the new century*. Washington, DC: National Academic Press.

New Media Consortium. (2014). *NMC horizon report: 2014 higher education preview*. Austin, TX: Author.

Redecker, C., Leis, M., Leendertse, M., Punie, Y., Gijsbers, G., Kirschner, P., Stoyanov, S., & Hoogveld, B. (2011). *The future of learning: Preparing for change*. Luxembourg: JRC European Union. Retrieved from http://ftp.jrc.es/EURdoc/JRC66836.pdf

Scott, G., & Yates, K. W. (2002). Using successful graduates to improve the quality of undergraduate engineering programmes. *European Journal of Engineering Education, 27*(4), 363–378.

Siemens, G. (2004). *Categories of e-learning*. Retrieved from http://www.elearnspace.org/Articles/elearningcategories.htm

Staron, M. (2011). Connecting and integrating life based and lifewide learning. In N. J. Jackson (Ed.), *Learning for a complex world: Lifewide concept of learning* (pp. 137–159). Bloomington, IN: AuthorHouse.

Stocklmayer, S., Rennie, L., & Gilbert, J. (2010). The roles of the formal and informal sectors in the provision of effective science education. *Studies in Science Education, 46*(1), 1–44.

Washington Accord. (2013). *Graduate attributes and professional competencies*. Retrieved from http://www.ieagreements.org

Willems, J., & Bateman, D. (2013). Facing up to it: blending formal and informal learning opportunities in higher education contexts. In G. Trentin & M. Repetto (Eds.), *Using network and mobile technology to bridge formal and informal learning* (pp. 93–118). Cambridge, UK: Woodhead.

Zimmerman, B. J. (2000). Attainment of self-regulation: a social cognitive perspective. In M. Boekaerts, P. Pintrich, & M. Zeidner (Eds.), *Self-regulation: Theory, research, and applications* (pp. 13–39). Orlando, FL: Academic.

Index

© Springer International Publishing AG, part of Springer Nature 2018
D. Corrigan et al. (eds.), *Navigating the Changing Landscape of Formal and Informal Science Learning Opportunities*,
https://doi.org/10.1007/978-3-319-89761-5

Printed in the United States
By Bookmasters